# The Union for the Mediterranean

This is the first comprehensive analysis of the Union for the Mediterranean (UfM), launched in 2008 amid great controversy within the European Union. Affected from the start by negative fallout from the failure of Middle East peace initiatives, its inadequacies have been underlined by the popular movement for regime change in the Arab world.

Leading experts provide here the first integrated analysis of the significance and shortcomings of the UfM. Beginning with critical questioning of the motives and institutional logics informing this venture, the collection proceeds to analyse its key actors, as well as major policy dossiers such as energy and development.

The book explains how and why an initiative aiming to depoliticize Euro-Mediterranean relations in fact proved wide open to political discord, bringing huge disruption to UfM activity. While some aspects are found to have merit, the volume is critical of the way in which EU Mediterranean policy became driven by a narrow range of national interests, lost sight of the political objectives of the preceding Barcelona Process and became overwhelmingly bilateral in approach, at the expense of more ambitious region-building efforts.

It concludes by highlighting the need to reform the EU Mediterranean policy framework in the light of the Arab uprisings of 2011.

This book was previously published as a special issue of *Mediterranean Politics*.

**Federica Bicchi** is Lecturer in International Relations of Europe at the London School of Economics.

**Richard Gillespie** is Professor of Politics at the University of Liverpool and founding editor of *Mediterranean Politics*.

# Acknowledgements

This publication stems from a collaboration based in the Europe in the World Centre, University of Liverpool and in the Centre for International Studies at the London School of Economics, where a highly productive workshop was held on 10 May 2010. This event was funded by a British Academy grant to Richard Gillespie for a project on 'The Union for the Mediterranean: Significance for the Barcelona Process' (SG-51979) and contributions from the Centre for International Studies, LSE, as well as Routledge, our publisher. We would also like to express our thanks to EU officials and diplomats from France and the UK for their active participation in the workshop, to the anonymous external reviewer of the collection for some very valuable feedback on an earlier draft, and to all contributors for bearing with us during several rounds of revisions of the papers: we learned a lot in the process and hope the readers will do so too.

Federica Bicchi

Richard Gillespie

# The Union for the Mediterranean

*Edited by*
Federica Bicchi and Richard Gillespie

LONDON AND NEW YORK

First published 2012 by Routledge

2 Park Square, Milton Park, Abingdon, Oxfordshire OX14 4RN

711 Third Avenue, New York, NY 10017

*Routledge is an imprint of the Taylor & Francis Group, an informa business*

First issued in paperback 2018

Copyright© 2012 Taylor & Francis

This book is a reproduction of *Mediterranean Politics*, vol. 16, issue 1. The Publisher requests to those authors who may be citing this book to state, also, the bibliographical details of the special issue on which the book was based.

All rights reserved. No part of this book may be reprinted or reproduced or utilised in any form or by any electronic, mechanical, or other means, now known or hereafter invented, including photocopying and recording, or in any information storage or retrieval system, without permission in writing from the publishers.

Notice:
Product or corporate names may be trademarks or registered trademarks, and are used only for identification and explanation without intent to infringe.

*British Library Cataloguing in Publication Data*
A catalogue record for this book is available from the British Library

ISBN13: 978-0-415-68964-9 (hbk)
ISBN13: 978-1-138-37724-0 (pbk)

Typeset in Times New Roman
by Taylor & Francis Books

**Disclaimer**
The publisher would like to make readers aware that the chapters in this book are referred to as articles as they had been in the special issue. The publisher accepts responsibility for any inconsistencies that may have arisen in the course of preparing this volume for print.

# Contents

1. The Union for the Mediterranean, or the Changing Context of Euro-Mediterranean Relations
   *Federica Bicchi* — 1

2. The UfM's Institutional Structure: Making Inroads towards 'Co-Ownership'?
   *Elisabeth Johansson-Nogués* — 19

3. France and the Union for the Mediterranean: Individualism versus Co-operation
   *Mireia Delgado* — 37

4. Adapting to French 'Leadership'? Spain's Role in the Union for the Mediterranean
   *Richard Gillespie* — 57

5. Germany and Central and Eastern European Countries: Laggards or Veto-Players?
   *Tobias Schumacher* — 77

6. The UfM and the Middle East 'Peace Process': An Unhappy Symbiosis
   *Rosemary Hollis* — 97

7. *Plus ça change. . .?* Israel, the EU and the Union for the Mediterranean
   *Raffaella A. Del Sarto* — 115

8. The Ties that do not Bind: The Union for the Mediterranean and the Future of Euro-Arab Relations
   *Oliver Schlumberger* — 133

9. A New Beginning? Does the Union for the Mediterranean Herald a New Functionalist Approach to Co-operation in the Region?
   *Patrick Holden* — 153

## CONTENTS

10. The UfM and Development Prospects in the Mediterranean: Making a Real Difference?
    *Diana Hunt*   169

11. Third Time Lucky? Euro-Mediterranean Energy Co-operation under the Union for the Mediterranean
    *Hakim Darbouche*   191

12. The UfM Found Wanting: European Responses to the Challenge of Regime Change in the Mediterranean
    *Richard Gillespie*   211

*Index*   225

# The Union for the Mediterranean, or the Changing Context of Euro-Mediterranean Relations

FEDERICA BICCHI
Department of International Relations, London School of Economics, UK

ABSTRACT *This contribution analyses the set of conditions that made the Union for the Mediterranean (UfM) possible, highlighting the change vis-à-vis the Euro-Mediterranean Partnership (EMP) and the European Neighbourhood Policy (ENP). First, it develops a conceptual framework for the analysis of the actors contributing or opposing the initiative, according to their attitude, motivation and resources invested in the process. Second, it examines the institutional logics that underpin the UfM. It suggests that the UfM was launched because a very small group cajoled an uninterested majority into yet another initiative for the Mediterranean. The outcome represents a shift away from regionalism as conceived in the EMP. At the same time, the Arab–Israeli conflict has politicized and disrupted the agenda of the UfM, as national interests have come to the fore and democracy and human rights have receded.*

The Union for the Mediterranean (UfM), launched by the French President Sarkozy on 13 July 2008, is the latest development in the history of Euro-Mediterranean relations. The creation of the EEC, which established a customs union among European states, early on posed the problem of how to relate to their southern neighbours in economic terms and then, as the Europeans endeavoured to find a common voice in foreign affairs, in political terms too. The Global Mediterranean Policy (1972), the Renewed Mediterranean Policy (1990), the Euro-Mediterranean Partnership (EMP) (1995), all embodied these attempts at finding a common platform for dealing with Mediterranean non-members.

The latest addition by Sarkozy introduces a set of novelties, the consequences of which are still unknown. It creates a co-presidency for the southern rim, while it institutionalizes meetings at the top level of heads of state and government, as well as a small Secretariat.[1] It emphasizes the partnership between the public and the private sectors. It stresses functional projects among bordering countries.[2]

It expands membership to include Bosnia-Herzegovina, Croatia, Montenegro, and Monaco. After the 'big bang' of the EMP, which expanded the number of issues on the agenda and the institutional setting of Euro-Mediterranean relations, the UfM thus recalibrates the balance by fine-tuning some key aspects.

In academic terms, this represents a challenge, as it raises a set of important questions. Why was the UfM launched? What conditions made it possible (and, for some, desirable)? What is the meaning of the changes it has introduced? What likely outcomes can we expect? These are the issues that this collection sets out to address. The aim in this contribution is to look at the broad framework, the specific parts of which will be the focus of the following contributions.

As with any political initiative, the UfM epitomizes a time-specific political context, which is bound to affect future developments. It is borne out of and interacting with the political agential inputs that key players in the area aim to impress onto the overall system, within a broader set of macro- and micro-trends. At the same time, it is bound to have an impact on future interactions. In this respect, the UfM is not exceptional, as various types of institutionalist analysis argue. As Riker put it (1980: 445), institutions are 'congealed' preferences. Whereas preferences tend to vary relatively often, the decision to establish an institutional framework crystallizes a set of preferences and a specific constellation of powers. The reach of those preferences and powers is thus extended across time. The institutional setting is therefore not only the symptom of its time, but it also affects the near future by creating the playing field within which new and possibly different preferences will come to have relevance. This argument is shared by scholars from a sociological perspective, who contest the micro-analysis of rational choice, but embrace the view of institutions as shared rules, practices and normative understandings that resist change (March and Olsen, 1989).

The academic challenge is to understand where the UfM comes from, what set of preferences, rules and practices it embodies and to which likely outcomes and appropriate behaviours it is going to lead. More specifically, the UfM must be understood in relation to the EMP and to the European Neighbourhood Policy (ENP), launched in 2004. While the UfM embodies a dynamics of its own, it was established in a thick institutional context. Therefore, while bringing a degree of novelty and rising out of a radically different political context, the UfM is expected to relate to an already well-established set of practices and roles.

There are elements of both continuity and change embodied in the UfM. Much can be said in favour of continuity. The country promoting the UfM was no surprise for Euro-Mediterranean relations. Once again, France championed the cause of the Mediterranean while making a case for France's international profile.[3] The processes that motivated the actors involved also seemed very much the same. Security, migration, energy, development, Arab–Israeli relations – all are issues the roots of which go back at times to the 1970s. They seem to remain the top priorities for all countries involved, regardless of the everlasting differences in approach. The southern Mediterranean countries welcomed the international attention that the initiative once again drew. Moreover, much of the fundamental structure of the EMP went untouched. At the core, the organizational setting remained the same, despite the addition of the Secretariat, co-presidency and top level meetings. The issues addressed in the multilateral discussion also continue to largely reflect the agenda of the EMP, although it could be argued that their normative value has changed.

Much can also be said in favour of change, regardless of the degree of apparent continuity. It cannot be assumed that an exuberant French president is all it takes to explain the new initiative, if only because the context of Euro-Mediterranean relations has substantially changed since the EMP was launched. The EU has undergone its biggest enlargement, nearly doubling in size. The shadow of enlargement was one of the triggers of the EMP (Barbé, 1998) and was thus somehow encompassed in the Euro-Mediterranean architecture of the 1990s. Most importantly, the nature of European integration seems to have subtly changed, and so has the EU agenda. The Franco-German integration engine has largely ground to a halt, leaving room for initiatives led by a small number of countries and, most crucially for our case, for French efforts to regain a leading role. The attack on multilateralism seems to have occurred in Euro-Mediterranean relations too, with a shift of emphasis in governance structures for co-operation. The existence of multilateral fora (a legacy of the EMP) seems to have lost relevance vis-à-vis the increase in bilateral relations (especially with the ENP) and notwithstanding the continuity of the EU unilateral financial instruments towards the area. The 'dialogue' about democracy and human rights has vanished. Moreover, in the Middle East there is no peace process to speak of. The Obama Administration faces a particularly hard-line Israeli Administration and no serious negotiations are in evidence. While falling short of a revolution, all these changes certainly represent a substantial evolution from the mid-1990s to now.

The argument presented here, which is to be read in dialogue with the following contributions, is that, despite appearances, change prevails over continuity. Although people not versed in the details of Euro-Mediterranean relations would be excused if they struggled to perceive a difference between the EMP and the UfM, this contribution will put forward the thesis that the UfM is the symptom of different political preferences on the part of the main actors and it is going to impress a different direction on Euro-Mediterranean relations, marking the UfM as a step in the fragmentation of an artificial region.

The following analysis focuses on actors and institutional logics, and on how the institutional order of Euro-Mediterranean relations has reflected a change in emphasis in these two dimensions. The first part focuses on the conceptual categories for the analysis of actors, in order to examine which actors have made the UfM possible and which have resisted it. The second part will address institutional logics, namely regionalism-bilateralism and functionalism-politicization, showing that the UfM reflects a weakened regionalism in the area (including within the EU) and displays a high degree of regional politicization, due to the collapsed Arab-Israeli peace process. The last part will bridge the analysed dimensions and compare them across time. It will show that, thanks to the entrepreneurial and/or leadership efforts of key actors, the institutional setting of Euro-Mediterranean relations has shifted from 'regionalism+politicization' in the EMP, to 'bilateralism+functionalism' with the ENP, to 'bilateralism+politicization' in the UfM.

## Actors

The focus on actors and the UfM raises the questions of who did it, why, by what means, and what role the other actors played or, to put it differently, what kind of dynamics emerged among so-called partners. The story of how the UfM came about has been told elsewhere (see Bauchard, 2008; Gillespie, 2008; Balfour, 2009). It is well established that France was in the driving seat in leading the initiative. While the UfM can be seen as a one man's effort, in the person of the French president, the dynamics that it engendered were much more complex than that. The issue arrived on the political agenda 'from above', as a result of high domestic politics. But the structure launched in 2008 differed from the early proposal by Sarkozy in 2007 in several respects, most importantly in terms of membership, which eventually included 43 countries (27 EU member states, 12 EMP partners on the southern Mediterranean rim and 4 new additions, Bosnia-Herzegovina, Croatia, Montenegro, Monaco). Moreover, very much like the run-up to the Barcelona Conference, the momentum behind the launching of the UfM developed 'first and foremost [as] an aspect of European foreign policy' (Gillespie, 2008: 278). But the preliminary interactions went beyond the borders of the EU. The reaction of the southern countries to the preliminary version of the UfM showed an increased determination to participate as full members in the new framework and criticisms tended to focus on the extent to which the new initiative would allow the full expression of such an intention.[4] It is thus important to scrutinize the role of the participant countries in bringing about the initiative, in order to forecast the potential for change of the UfM.

There are three characteristics that are useful in identifying the role that various actors played in bringing about the UfM (and policy initiatives in general): attitude, motivation and amount of resources invested. The attitude of actors is the first step in analysing the dynamics of agenda setting and decision making. Did actors support the initiative or did they try to resist it? In general terms, we can distinguish actors among leaders, laggards, and fence-sitting actors. According to the amount of resources invested, leaders playing a central role *against* an initiative can, however, act as veto-players, blocking its adoption. Moreover, the motivation of leaders helps to distinguish between, on the one hand, strategic leaders and, on the other, genuine entrepreneurs that strive to achieve consensus in the name of the common good. Finally, marginal players might behave as low-profile supporters or unhappy laggards, but they can also strategically look for side payments in exchange for their support or collectively block developments through lack of enthusiasm.

While France obviously supported the initiative and can be identified as the main actor behind it,[5] the other key actor was Germany, though not in its traditional role. France put the issue of the Mediterranean on the EU agenda in an indirect way, as the *Union Méditerranéenne* (UM) was sketched out to a domestic audience, by a yet-to-be-elected candidate for the Presidency. Once the elections were over, the new president did not involve the EU and on the contrary continued to work on a proposal that would have marginalized it. Germany's reaction was fence-sitting at first, and then 'calling the bluff' by acting as a veto-player.[6] Based on the old saying of 'no taxation without representation', Germany's role was pivotal in bringing about substantial changes to the initiative and in establishing a role for the EU. While the amount of material resources invested by Germany in the endeavour was not high, the political capital invested in facing Sarkozy was substantial and very public, although it was not alone. In fact, Germany spearheaded a group of countries that preferred the involvement of the entire EU and the continuation of the EMP in a different guise. This silent majority was composed not only of northern European countries, but also of Arab ones (see Driss, 2009: 2; Kausch and Youngs, 2009: 963; Schlumberger, this collection). These countries were unhappy laggards, which at times played fence-sitting and waited for Germany to take the lead in suggesting/imposing reforms to the initial project.[7]

Since the shift from the *Union Méditerranéenne* to the UfM, central and eastern European countries oscillated between being low profile supporters, favour exchangers and unhappy laggards, calling for an eastern equivalent and thus supporting the Eastern Partnership (see Schumacher, this collection). Other northern European countries, such as the UK, maintain a low profile on the issue, reflecting the low priority assigned to the dossier and the lack of interest in what is regarded as an essentially French political game.

Spain and Italy tried to work as co-entrepreneurs,[8] but they met with the determination with which France tried to establish itself as the sole leader. This pattern broke with the co-operation that had emerged between France and Spain in the run-up to the Barcelona Conference, and it was instead inspired by previous forms of co-operation. In the case of the EMP, Spain invested a great deal of political capital in promoting the initiative from the early 1990s, but it was ready to co-ordinate with France, which since 1994 and until the Barcelona conference behaved as a de facto co-entrepreneur with Spain (Gillespie, 1997: 38). The run-up to the launch of the UfM was instead a very French endeavour, which resembled French behaviour leading to the GMP in 1972. At the time, capitalizing on the ongoing discussions about the role of Mediterranean countries during a period of détente, France outmanoeuvred other proposals on the table to promote the first EEC initiative towards the Mediterranean (Bicchi, 2007: 91–7). Similar to the UfM, the French activism entailed a number of 'surprises' for its European partners before they reached a common decision on the GMP.[9] In 2008, France did not limit surprises for its partners to the issue of the Mediterranean (Schwarzer, 2008: 366), although the lack of communication on this dossier represented a major breach to the Common Foreign and Security Policy (CFSP)'s plea for solidarity among member states on matters of foreign policy. Southern European partners were thus relegated to the role of low-profile supporter or favour exchanger, despite the amount of resources poured into the issue, as has been the case for Spain.

Turkey and, to a lesser extent, Israel were the countries whose attitudes remained consistently (although not vocally) negative about the new endeavour. From the point of view of Turkey, a central role in the Mediterranean could not in any way compensate for the lack of a role in Europe (Schmid, 2008). At the same time, Turkey's attitude was also lukewarm towards the EMP; yet, despite that, Turkey has been very active in negotiating free trade agreements with southern Mediterranean countries along the lines of the Euro-Med Association Agreements. Although this is a legal requirement related to Turkey's customs union with the EU in 1995, Turkey has embraced the endeavour with both an economic and a political interest, at times succeeding where the EU has failed, as in the case of Syria or Georgia.[10] The attitude is thus negative but functional and conditional on the achievement of some tangible benefits. Israel is a similar case (see Del Sarto, this collection). Having largely benefited from the increased bilateralism embodied in the ENP, any return to a multilateral forum detracted from the status quo. However, while France represented for Turkey an obstacle on the path to full membership, Israel perceived France (and more specifically Sarkozy) as a crucial ally in relations with the EU. Both countries thus were laggards that needed to be bought off at specific moments in time in order to become favour exchangers instead of veto-players.

The picture that emerges from this analysis thus suggests that the UfM developed as the outcome of the efforts of a small number of countries. France, supported at its discretion to by Spain and Italy, accepted a crucial change in the original plan in order to achieve the acquiescence of a large set of countries, represented by Germany, which favoured more continuity with the EMP than in the original plan. Since this fundamental compromise, the history of the UfM has comprised a set of small compromises to buy off a number of strategic but relatively marginal players against a background largely of indifference to or disillusion with political change in Euro-Mediterranean relations. To put it in politically incorrect terms, the UfM was launched because a very small group cajoled an uninterested majority into yet another initiative for the Mediterranean. However, now that potential dissenters have been bought off with side payments, a majority of participants have a stake in the project, though generally small and potentially counterproductive for the common good of Euro-Mediterranean relations. As a consequence, the UfM can rely on a limited amount of political capital in case of difficulties, because it represents different things to different actors.

## Institutional Logics[11]

The institutional architecture embodied by the UfM is another aspect worth considering in detail. Every institutional design expresses a political plan. In relation to the Euro-Mediterranean organizational context within which it is situated, the UfM represents a shift of emphasis in two key dichotomies: regionalism/bilateralism, and functionalism/politicization. The UfM represents a shift away from regionalism as conceived in the EMP and a further weakening of the region-building strategy of the EU in the Mediterranean. At the same time, the expected depoliticization of the regional dimension of Euro-Mediterranean relations is unlikely to take place.

### *Regionalism/Bilateralism*

The regionalist strategy of the EU[12] has a long, dual history (Bicchi, 2007). It started with the GMP, when member states 'invented' the Mediterranean as a political area that was homogenous enough to justify addressing all parts in the same way. The GMP consisted in nearly identical, parallel bilateral channels, with no multilateral framework however. One of the main novelties of the EMP (if not the main one) was rather the degree of regionalism embedded in the endeavour and the multilateral setting it created. The EMP thus set out to 'construct' the Mediterranean, by establishing a semi-permanent, multilateral dialogue on a very broad agenda, as indicated by the three baskets of the Barcelona Declaration. Faced with a number of perceived security issues, the EU addressed them by region-building, in the form of regional dialogues, rather than by intensifying intra-European security co-operation (Adler and Crawford, 2006). The extent to which this was done with the final goal to create a common Euro-Mediterranean region, rather than a separate non-European Mediterranean region (e.g. Pace, 2006) is a matter for discussion. The nature of the relationship often corresponded 'more to a soft form of hegemony than to a partnership' (Philippart, 2003: 215), largely reflecting the imbalance in terms of economic and political power (cf. Holden 2009). Nevertheless, the institutionalization of the EMP's multilateral dimension was an undeniable achievement in comparison with the former 20+ years of Euro-Mediterranean relations.

The ENP, on the contrary, re-introduced a strong degree of bilateralism (Del Sarto and Schumacher, 2005). The 'regatta approach', the granting of 'advanced status' to selected partner(s) and the negotiation of agreements in addition to the Euro-Med Association Agreements signalled that the relationship between the multilateral dialogue among all participants and the agenda for bilateral relations was reversed: rather than the multilateral dialogue setting the themes to be then adopted and adapted in bilateral relations, bilateral relations were to explore avenues that could not be addressed at the multilateral level. Only one indication remained of the ambitious plan for a Euro-Mediterranean free trade area, namely the pan-Euro-Mediterranean protocol on rules of origins, which does contribute to the original goal, but in a much less demanding way.

The Union for the Mediterranean represents a further step towards bilateralism and away from regionalism, in at least two ways. First, a key aim of UfM is to promote projects among groups of willing countries (see Darbouche, this collection; Hunt, this collection), especially in geographically contiguous areas. Potential for sub-regional co-operation certainly exists (see Darbouche, this collection). Moreover, the shift of emphasis can be depicted as a sign of pragmatism (Seeberg, 2010), attracting rather than coercing countries into co-operation. At the same time, the importance assigned to the sub-regional level implicitly recognizes that the regional level cannot deliver. Its institutionalization sets an order of preferences different from the all-inclusive, multilateral setting. It stresses coalitions of the willing based on functional complementarities or overlapping visions more than it promotes ambitious plans to create common political projects out of dissent. In short, it downsizes the political significance of EU foreign policy towards the area, although it also introduces a degree of realism.

Second, the increase in the number of participants further contributes to the dilution of regionalism. If consensus in the EMP was difficult, the addition of three south-eastern European countries is not going to make it easier. On the contrary, by increasing the range of diverse interests that must be accommodated, it amplifies the need to focus on sub-regional projects and the related impossibility to achieve anything substantial with over 40 members.

The diminished emphasis on regionalism and multilateralism is not limited to Euro-Mediterranean relations. On the contrary, it very much characterizes intra-EU relations on matters of foreign affairs, as best exemplified by the substance of the UfM and by the way in which EU member states came to an agreement on the UfM. The matter focused on different visions about the extent to which Europe and the EU should have featured in the new initiative, which sparked an intergovernmental discussion among member states. While the matter was debated at EU meetings, the main decisions were not taken therein, but rather at the national level or in bilateral contacts. The French proponents of the UM looked to the EU to provide a large proportion of the mixed funding plans for the new initiative, but their thinking was not about how to work through the EU framework to bring about the new initiative. Rather, France sought to offset German pre-eminence in the EU by thwarting attempts to adopt common EU positions in relation to the Mediterranean, while getting its UM initiative off the ground through selective unilateral approaches at the bilateral level, initially primarily to the Mediterranean countries.

Since the formal launch of the UfM, all participants have more or less readily accommodated an intergovernmental(ist) approach to the UfM (Gillespie, forthcoming), which contributes to the fragmentation of the multilateral logic previously embedded in the EMP. France has continued to impress an intergovernmental character on the debate, by managing to extend its co-presidency of the UfM to two years. It did so by persuading the Czech Republic, Sweden (reluctantly) and Spain to surrender their own rights derived from their successive EU presidencies.[13] Italy has also embraced an intergovernmentalist approach under its traditional 'European' discourse (Tassinari and Holm, 2010: 15–16). More generally, this chimes well with the nature of discussions within the CFSP, which since the 2004 enlargement have emphasized the relevance of minilateral gatherings, with a small group of countries co-ordinating their actions with a view to affecting the negotiations between the 27.[14] Southern Mediterranean countries have happily followed the new tune, which suits their geopolitical strategies much better.

In the light of the above, the discussion about the extent to which the UfM reinforces rather than replaces the EMP is anodyne, as the UfM follows from the EMP but it fundamentally changes one (though not just one) of its main aspects. While the UfM de facto continues the EMP, its structures and its agenda, it has scaled back its multilateral component. The substance of Euro-Mediterranean cooperation is thus no longer bloc to bloc (EU+Med) as in the EMP, or bloc to single country (EU+single Med countries) as in the ENP, but single country to single country.

The 'convergence of civilization', in favour of which Adler and Crawford (2006) have argued, has thus suffered a setback because of the weakened region-building strategy of the EU.[15] This development is related not only to developments in the Arab–Israeli peace process (on which more below), but also to the intra-European fragmentation. How lasting the consequences of this shift are going to be remains to be seen, although at the moment it is fair to agree with Kausch and Youngs (2009: 963) that the legacy of the EMP is 'on life-support'.

## Functionalism/Politicization

The second fundamental dichotomy in the analysis of Euro-Mediterranean relations is the juxtaposition of functionalism and politicization. Much of this dichotomy reflects the low/high politics spectrum (see Holden, this collection). In this respect, the UfM displays a complex pattern. It was marketed as a 'relaunch' of Euro-Mediterranean relations, which would rescue it from the creeping politicization of all dossiers. However, at the moment it is questionable whether it has delivered an increase in the functional logic. At the same time, it has so far decreased the overall political significance of Euro-Mediterranean relations, while being unable to resist the creeping ascent of national interests. In the shift from a more regionalist to a more bilateral and intergovernmentalist approach, the level of politicization of technical dossiers has remained relatively high, thanks to the collapse of the Arab–Israeli peace process, although preferences institutionalized in the new structure are more parochial and partisan than before.

At first glance, there is an increase in functionalism.[16] The UfM has been defined as 'a union of projects' (or, more to the point, 'a project of projects'). In this respect, it represents a number of ambitious innovations, leading to the creation of specialized technical agencies as well as new political and administrative institutions. It also calls for a new partnership between public and private, especially in the financing of the new projects, including the possibility of an involvement of capital from the Gulf countries and the creation of a new financial instrument, possibly funded by creating a subsidiary of the European Investment Bank.[17] The new projects would however occur in an area in which there is already substantial activity, much of which was set in motion by the EMP (Emerson, 2008). It is thus difficult to see the UfM as a genuine 'opportunity' to introduce more enlightened policies in this respect, as diplomatic circles in Paris suggest.

Moreover, apart from the case of solar energy (see Darbouche, this collection), the current situation does not suggest how opportunities might turn into concrete achievements, as little interest has so far emerged on several dossiers and rumours abound about the potential bankruptcy of the UfM Secretariat. A test of commitment will be whether southern Mediterranean countries start to propose new fields of action for future UfM projects. Thus far the project proposals have come mostly from France and then have been modified and reduced in number on the basis of negotiation with interested EU member states and Commission officials.

While functionalism might not become the new foundation stone of Euro-Mediterranean relations, politicization at the regional level of dossiers remains high, although in a different form than in the EMP. At present, we can distinguish three separate processes affecting the politicization of the UfM: 1) the Arab–Israeli conflict has come to affect the schedule of and the topics on the Euro-Med agenda, 2) national interests, promoting country-specific issues, have also come to the fore, 3) good governance and human rights have descended in the list of priorities.

First, the most prominent driver of politicization in Euro-Mediterranean relations is the Arab–Israeli conflict. The link between Euro-Mediterranean relations and the Middle East (and Arab–Israeli relations more specifically) exists for a number of reasons: the symbiotic relationship that many (especially in the EU) saw between the Barcelona Process and the Middle East peace process in the past; the direct involvement of several Mediterranean partners in the conflict; the wider resonance of the Israeli–Palestinian conflict in the Arab countries and in Europe; and the fact that, among several conflicts around the Mediterranean Basin, it remains a 'hot' conflict, rather than a frozen one. The UfM came into being without any contemporaneous progress in diplomatic peace-making efforts between Israel and the Arab countries, although diplomatic relations between Syria and Lebanon were being normalized at the time. This was a development showcased at the Paris summit in 2008, but which in fact has not shown any progress since.

We are now witnessing a nearly complete symbiosis between the UfM and the Arab–Israeli conflict (see Hollis, this collection). The merger has occurred slowly but relentlessly, as demonstrated by the increasing relevance of the fallout from Arab–Israeli relations on the EMP/UfM agenda, helped by the increasing importance of parochial national interests. The sequence is worth looking at in detail[18] in order to fully appreciate the growing impact of the Arab–Israeli conflict and the parallel slowdown of the proceedings.

The first very public halt to the EMP occurred in 2005, when the first top level summit, organized by the British Presidency of the EU and Spain to mark 10 years since the Barcelona Conference, failed at the last minute when all the southern Mediterranean heads of state and government, apart from Turkey and the Palestinian Authority, failed to show up in a loosely organized protest against the perceived pushiness of the EU in promoting the 'fourth basket' (migration and internal security). This had little direct link to Arab–Israeli relations. Prior to 2005, however, Lebanon and Syria had already boycotted ministerial meetings in Egypt and in Morocco because of Israel's official presence in an Arab country, as well as sending low level representations at times of tension on the Arab–Israeli front.

Interestingly, the war in Lebanon in 2006 did not halt the EMP proceedings. Arab countries threatened to boycott meetings, but the Finnish Presidency of the EU at the time made it clear that meetings would take place whether participants liked it or not. It thus exerted a steering role that it will be impossible to achieve under the UfM co-presidency without the co-operation of the co-president (unlikely, in such an occurrence).

The ink on the UfM Declaration was not yet dry when the working schedule ground to a halt because of Israel's opposition to the incorporation of the Arab League as an observer with rights to intervene (see Del Sarto, this collection). While it had become customary for a representative of the Arab League to participate in EMP meetings, the UfM offered the opportunity to formalize this participation. Participant countries found an agreement in that sense, as expressed in a mention in the Declaration issued at the Paris summit in July 2008. However, when the Arab League representative made a statement at the very first meeting in September 2008, Israel objected and argued that in its understanding, the Paris Declaration did not grant the right to speak to the Arab League representative. The issue brought meetings to a stop until November 2008. Shortly afterwards, the Gaza war, which spanned across December 2008–January 2009, led this time to the Arab boycott of all meetings, which was protracted until July 2009, when meetings were restarted first at the very low end of the hierarchy and progressively climbing to higher levels. The election of Netanyahu further complicated things during this period, as it hardened both sides, with some Arab countries objecting to a restart of the proceedings with an Israeli government that had not yet recognized the two states solution.

As soon as low level meetings had resumed in summer 2009, negotiations suffered a further setback. The personal row between the Egyptian and the Israeli ministers of foreign affairs escalated to the point of Egypt declaring a boycott of the ministerial meeting in Istanbul in November 2009. Here too, the institution of the co-presidency worked against a common political Euro-Mediterranean project, as Egypt called for the cancellation of the meeting in its role of co-president, claiming that it represented the position of all Arab countries, whereas North African countries did not see eye to eye with Egypt on this occasion. Turkey being the host of the cancelled meeting made it easier for Egypt to disrupt the UfM proceedings.

2010 did not start off any better and ended even worse. Three sectoral meetings (devoted to water,[19] tourism[20] and agriculture[21]) were hampered because of the Arab–Israeli conflict. Despite having reached a consensus on the working programme, the ministerial meeting on water failed to issue a final declaration because participants could not agree on how to refer to the 'occupied territories'. The difficulties in organizing the second summit bringing together heads of state and government soon became overwhelming, despite the efforts of Spain and of Moratinos in particular. In order to avoid clashes among foreign ministers, it was envisaged that the preparatory meeting might be held at the level of senior officials instead. However, the photo opportunity for all sides in the Arab–Israeli conflict was not justified by the results of the resumed proximity talks. There was very little chance of glory, and this (together with the thorny issue of the European co-presidency) affected the commitment of France to a summit in June. Spain had thus to accept a postponement to November 2010, although the venue would still be Barcelona. France tried to separate the Middle East from the UfM by suggesting a conference on the former in Paris at the beginning of November, followed by the UfM summit. But given the breakdown in negotiations after the resumption of construction of Israeli settlements in September 2010, both events were cancelled and, at the time of writing,[22] no further date has been set for the UfM summit.

This timeline thus shows that while the EMP has encountered some substantial difficulties in the last few years of its existence, meetings in the UfM have been affected by the Arab–Israeli conflict to an unprecedented degree. Since the creation of the UfM, the calendar of meetings has hardly worked without disruption. In fact, there were barely any low level meetings in the first year of the UfM existence, while there have been no meetings of foreign ministers since the end of the French Presidency in 2008, not to mention the absence of further summits. Even sectoral meetings have now fallen prey to the Arab–Israeli conflict. It is thus ironic that some critics of the Barcelona Process have argued that it was too closely linked to the Middle East peace process initiated by the Oslo Agreements and that, in order to succeed in relation to regional development objectives, a new initiative was needed.

As the recent experience shows, the new institutional framework has not escaped the politicization of the issue. On the contrary, it has witnessed the near complete overlapping of agendas.

There is a second, less prominent but equally substantial, strand of politicization in the UfM, which relates to national interests. The UfM is vulnerable to fallout from an increased number of regional conflicts and disputes, now that the Partnership has expanded in the western Balkans. Tensions between Turkey, Greece and Cyprus also affected UfM activity in the first half of 2009, as Greece and Cyprus opposed Turkey being given a deputy secretary-general post in the UfM. The opposition faded when Turkey in exchange agreed to drop its standard opposition to Cyprus holding international posts. Cyprus, Greece and Turkey would have agreed to resolve their differences without blocking negotiations within the UfM, though. The whole process of the choice of deputy secretary-generals was a chapter in the increasing horse-trading techniques that participants were exerting in the definition of institutional details. Jordan secured the secretary-general position early on in the dispute, while the number of deputies, which eventually climbed to six, testifies to the various trade-offs (see Johansson-Nogués, this collection). Moreover, northern European countries kept themselves out of the dispute, and in fact of several other roles as participants, such as in the creation of the financial institution in which the Germans and the British, among others, have declined to participate.

These politicization processes parallel an equally strong third process, which works in the opposite direction. The UfM has marked the depoliticization of one of the very few progressive chapters in Euro-Mediterranean relations, namely human rights and good governance. As argued in the literature, the emphasis is shifting from good governance to 'good enough governance' (Kausch and Youngs, 2009: 974; Tassinari and Holms, 2010), which stems from 'pragmatic' considerations (Seeberg, 2010) and substantial downscaling of ambitions in this field. The political project of 'constructing a Mediterranean region' based on democracy and human rights has been largely abandoned amid a progressive fragmentation of efforts.

The overall balance of the UfM on the dimension ranging from functionalism to politicization is simultaneously an increase in the politicization of Euro-Mediterranean relations *and* a step in the direction of depoliticization.[23] It is highly politicized at the regional level, because of the Arab–Israeli conflict, while at the same time it is depoliticized in its content, because of the low interest in any project of political transformation. Paradoxically, but not so much so given the nature of internal Arab politics, the high politicization of Arab–Israeli relations is instrumental to the depoliticization of an agenda for domestic change: the higher the Arab–Israeli conflict remains in the attention of European and Arab audiences, the less scrutiny Arab rulers have to endure.

Therefore, the overall picture of the UfM in terms of institutional logics is complex. Regionalism has lost its appeal, while intergovernmental, bilateral relations have gained in relevance, both within the EU and in Euro-Mediterranean relations (Mediterranean countries never having been big fans of regionalism). Whereas a functionalist shift might take place, and if so only in the long run, the expected decline in politicization of the Arab–Israeli conflict did not occur. On the contrary, it took a turn for the worse, with near complete overlapping between the two agendas. The Barcelona Conference declared the EMP and the Middle East peace process 'separate but complementary'. The relationship between the UfM and the Arab–Israeli conflict can instead be described as 'overlapping and contradictory'. In parallel, transformative projects of the international context in the Mediterranean and of the domestic context of Mediterranean countries have lost their urgency and have slipped down the political agenda.

## From the EMP to the ENP to the UfM: The Evolution of Euro-Mediterranean Relations in the Post-Cold War Order

What is the evolution of Euro-Mediterranean relations since the end of the Cold War? The factors highlighted so far contribute to illuminate the trajectory from the EMP to the UfM, based on the assumption mentioned above that institutions embody actors' preferences at specific moments in time, capturing the *Zeitgeist*-like pictures. If we cross the two institutional dimensions analysed above and assume that events ranging from entrepreneurial actors to the state of Middle East conflicts and securitization of EU policies drive the institutionalization process,[24] we come up with a picture that (in a broad generalization) resembles the one depicted in Figure 1.

The masterpiece of the post-Cold War context was the EMP.[25] People launching the EMP captured the spirit of the time by thinking the previously unthinkable: the EU responded to newly perceived security threats with a highly political and innovative project to create a Euro-Mediterranean region (see Adler and Crawford 2006). It was a truly regional framework, for the first time in Euro-Mediterranean relations. And it was a very political initiative too, based on the intention by EU member states in particular to transform relations with and within their southern neighbours, the former being the precondition for the latter. The extent to which economic relations were the main driver of the process can be discussed. There is little doubt however that even the economic projects embodied by the EMP (most notably the Euro-Med FTA and the Euro-Med Association Agreements) had a highly political flavour due to their scope and breadth.

Fast forward a decade and in comes the ENP, with a very different institutional setting. The emphasis here is on bilateral relations, with no reference to regional

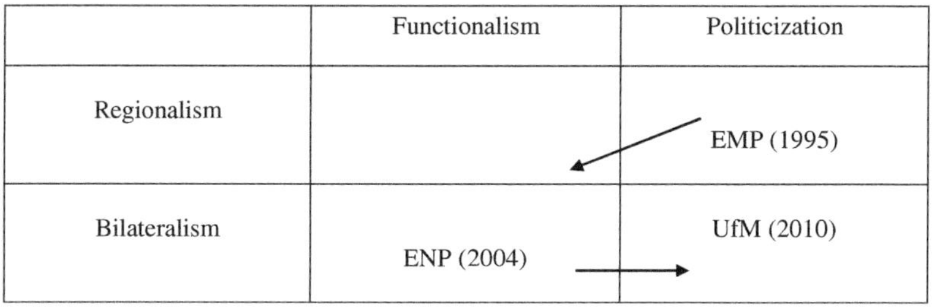

|  | Functionalism | Politicization |
|---|---|---|
| Regionalism |  | EMP (1995) |
| Bilateralism | ENP (2004) | UfM (2010) |

**Figure 1.** The institutional logics of the EU initiatives towards its Southern neighbours.

aspects including all members beyond a generic reference to the 'neighbourhood'. Moreover, the official discourse, as well as the daily practice, is imbued with references to specific projects and technical agreements. The common (bilateral) dialogues centre on the 'management' of the Association Agreements. There is a discussion about political issues, including political developments inside Mediterranean countries, but its value is limited and unconnected to the rest of the negotiations.

The UfM represents a further development. It is possible to argue that the original French project, which involved just countries bordering the Mediterranean Sea, was a regional one, although not centred on a Euro-Mediterranean region. Moreover, part of France's motivation for promoting a new institutional structure was to shift the focus away from the highly contentious and politicized issue of Middle East relations. In its early UM formulation, therefore, the project should have fallen into the empty cell in the matrix in Figure 1, combining regionalism with functionalism. However, as this contribution has argued, the context within which the project was discussed did not allow for such an outcome and pushed the UfM into the 'bilateralism+politicization' category.

Despite its post-Cold War setting, the UfM, more than the EMP did, thus marks the current difficulties or even the impossibility to go beyond the challenges that have characterized Euro-Mediterranean relations since the end of World War II. In that respect, the decision by France to shake the institutional order of Euro-Mediterranean relations at a moment of low intra-EU and intra-Euro-Mediterranean co-operation was not helpful, as even if co-operation increases in the future, this institutional structure will continue to cast its shadow.[26]

## Conclusion

The institutional setting created in July 2008 represents a change not only in the name but also in the substance of Euro-Mediterranean relations, although the ultimate extent of its impact is still unclear.

The UfM developed as the outcome of the efforts of a minority of countries cajoling a majority into accepting a new initiative for the Mediterranean. France, supported by Spain and Italy when necessary, led the initiative, though at first restricted just to riverain countries. However, it had to accept the active involvement of the EU in order to achieve the acquiescence of Germany and of a broad set of other countries from northern and eastern Europe, as well as Arab countries, favouring more continuity with the EMP than in the original plan. Israel and, even more so, Turkey have been lukewarm supporters of the initiative, as it diluted their special relationship with the EU. Dissent has led to side payments in order to avoid laggards becoming veto-players. Given the ad hoc nature of this agential coalition, the UfM has come to represent different things to different actors, with a limited political capital at its core.

The institutional logics embedded in the new structure mark a further shift away from the region-building strategy of the EU, which characterized the EMP. As the emphasis falls on sub-regional projects and the participants' number increases, the political project of creating a region in the area is diluted and political ambitions downsized. Contrary to expectations, the Arab–Israeli conflict, which has loomed over the last few years of the EMP, is now in near complete symbiosis with the UfM, influencing the pace of proceedings as well as the substance of negotiations. The dialogue on democracy and human rights is silenced. The UfM thus displays a complex pattern of politicization and de-politicization, while the functional aspects have yet to come into fruition.

It would be pointless to be nostalgic regarding the EMP. The UfM has come about because of and reflecting a different constellation of preferences. Since the launch of the ENP, the political context around the Mediterranean has changed, both in the EU, with enlargement, and in the Middle East, with a hardening of positions around the Arab–Israeli conflict. In the last few years, the conflict has been more intractable than ever and counters any region-building attempts in the area. At the same time, the EU has not been particularly ambitious in its foreign policy and it is still digesting the effects of enlargement and of the Lisbon Treaty. As the game got tough, the EU members have not been among the toughest joining the play, at least in Euro-Mediterranean relations. The main criticism that can be levelled against the UfM is that it was created at a time when no common discourse of the Mediterranean really existed, not even in the EU, and therefore the UfM, because of the nature of institutions, has 'congealed' this set of preferences. Even if the context undergoes positive changes, including in the Middle East, the UfM might not be able to quickly take them on board.

One issue thus remains open: if this is the state of affairs, what can be achieved in the area via intergovernmental, sub-regional or bilateral, co-operation?[27] Are we going to witness an exponential growth of small projects, which ideally would combine to support a functionalist approach to Euro-Mediterranean relations? Or the little projects agreed will not be cumulative or even point in different directions? Or, finally, is the UfM the kiss of death for Euro-Mediterranean relations, doomed to oblivion? We are living in interesting times, but for reasons that profoundly differ from those highlighted after the creation of the EMP.

## Acknowledgements

I would like to thank Richard Gillespie, whose idea it was to have an academic discussion and a special issue of *Mediterranean Politics* on the UfM and who co-drafted an earlier version of this paper, as well as commenting extensively on further versions. I am also very grateful for comments from the participants at the workshop 'The Union for the Mediterranean: Continuity or Change in Euro-Mediterranean Relations?' (London School of Economics, 10 May 2010) and to the participants at the workshop organised by Ulla Holm and Fabrizio Tassinari on Euro-Mediterranean relations (Danish Institute for International Studies, Copenhagen, 4 June 2010). I also thank Michelle Pace for written comments and an anonymous (and very fair) reviewer. The usual disclaimers apply.

## Notes

[1] See Johansson-Nogués (this collection) on the novelties introduced by the UfM and its institutional structure.
[2] For energy co-operation, see Darbouche (this collection).
[3] On French entrepreneurship, see Delgado (this collection).
[4] On the reaction of Southern countries, see Del Sarto and Schlumberger (this collection).

5 On France, see Delgado (this collection).
6 As Tobias Schumacher puts it (in this collection).
7 Interview, official in Germany's Permanent Representation in Brussels, May 2008.
8 On Spain, see Gillespie (this collection).
9 Including, arguably, in the run-up to the Euro-Arab Dialogue in 1973–74 (Bicchi, 2007: 102–103).
10 I would like to thank Serah Kekec for this point.
11 This part benefits from an earlier draft written jointly together with Richard Gillespie.
12 See Holden, this collection, for a discussion of the terms regionalism and regionalization.
13 At the time of writing, it is not yet clear how the controversy is going to be resolved within the EU. The most likely scenario is that the newly established European External Action Service (EEAS) is to replace country representation in the European co-presidency of the UfM. However, it remains to be seen whether the high representative/vice president will design a role for member states e.g. for deputizing to the high representative/vice president.
14 The Common Security and Defence Policy is a case in point.
15 I would like to thank Stefania Panebianco for raising this point.
16 For a more thorough analysis of functional aspects in the UfM, their significance and limitations, see Holden, this collection. See also the contributions by Darbouche and Hunt.
17 See *The Financing of Co-Development in the Mediterranean*, Final Report of the High-Level Working Group chaired by Mr Charles Milhaud, May 2010. Available at: http://www.economie.gouv.fr/directions_services/dgtpe/publi/rap_milhaud1009_en.pdf (accessed 18 October 2010).
18 I would like to thank an official in the General Secretariat of the Council of the EU for contributing to fine-tune the timeline in a phone interview, 15 July 2010.
19 It was held in Barcelona, 13 April 2010.
20 It was held in Barcelona, 20 May 2010.
21 It was due to be held in Cairo 15–16 June 2010.
22 December 2010.
23 For the depoliticization side of the argument, see Kausch and Youngs (2009); Seeberg (2010); and Schlumberger (this collection).
24 On the interaction between entrepreneurs and windows of opportunities, see Bicchi (2007).
25 Despite the fact that strictly speaking, it was not the first post-Cold War initiative, the Renewed Mediterranean Policy being the one.
26 It casts a shadow also on arguments about Europe as a 'force for good' or as a 'normative power' (Manners, 2002). (See also Schlumberger, this collection.).
27 I would like to thank Fabrizio Tassinari for raising this crucial question.

## References

Adler, E. & Crawford, B. (2006) Normative power: the European practice of region-building and the case of the Euro-Mediterranean Partnership, in: E. Adler, F. Bicchi, B. Crawford & R. A. Del Sarto (2006) *The Convergence of Civilizations: Constructing a Mediterranean Region* (Toronto, Buffalo, NY, and London: University of Toronto Press), pp. 3–47.

Balfour, R. (2009) The transformation of the Union for the Mediterranean, *Mediterranean Politics*, 14(1), pp. 99–105.

Barbé, E. (1998) Balancing Europe's eastern and southern dimensions, in: J. Zielonka (Ed.) *Paradoxes of European Foreign Policy* (Leiden: Kluwer).

Bauchard, D. (2008) L'Union pour la Méditerranée: un défi européen, *Politique étrangère*, 1, pp. 51–64.

Bicchi, F. (2007) *European Foreign Policy Making toward the Mediterranean* (New York and Basingstoke: Palgrave Macmillan).

Del Sarto, R. & Schumacher, T. (2005) From the EMP to ENP: what's at stake with the European Neighbourhood Policy towards the southern Mediterranean? *European Foreign Affairs Review*, 10(1), pp. 17–38.

Driss, A. (2009) Southern perceptions about the Union for the Mediterranean, *EuroMeSco Paper*, produced the Instytut Spraw Publicznych/Institute of Public Affairs. Available at http://www.

euromesco.net/index.php?option=com_content&task=view&id=1201&Itemid=53&lang=en (accessed 29 November 2010).

Emerson, M. (2008) Making sense of Sarkozy's Union for the Mediterranean, CEPS Policy Brief, 155 (Brussels: Centre for European Policy Studies).

Gillespie, R. (1997) Spanish Protagonismo and the Euro-Med partnership Initiative, *Mediterranean Politics*, 2(1), pp. 33–48.

Gillespie, R. (2008) A 'Union for the Mediterranean'... or for the EU? *Mediterranean Politics*, 13(2), pp. 277–286.

Gillespie, R. (forthcoming) The Union for the Mediterranean: an intergovernmentalist challenge for the EU? *Journal of Common Market Studies*.

Holden, P. (2009) *In Search of Structural Power: EU Aid Policy as a Global Political Instrument* (Farnham and Burlington: Ashgate).

Kausch, K. & Youngs, R. (2009) The end of the 'Euro-Mediterranean vision', *International Affairs*, 85(5), pp. 963–975.

Manners, I. (2002) Normative power Europe: a contradiction in terms? *Journal of Common Market Studies*, 40(2), pp. 235–258.

March, J. G. & Olsen, J. P. (1989) *Rediscovering Institutions. The Organizational Basis of Politics* (London and New York: Macmillan and Free Press).

Pace, M. (2006) *The Politics of Regional Identity. Meddling with the Mediterranean* (London and New York: Routledge).

Philippart, E. (2003) The Euro-Mediterranean Partnership: a critical evaluation of an ambitious scheme, *European Foreign Affairs Review*, 8, pp. 201–220.

Riker, W. H. (1980) Implications from the disequilibrium of majority rule for the study of institutions, *The American Political Science Review*, 74(2), pp. 432–446.

Schmid, D. (2008) La Turquie et l'Union pour la Méditerranée: un partenariat calculé, *Politique étrangère*, 1, pp. 65–76.

Schwarzer, D. (2008) La présidence française de l'Union européenne: quels objectifs, quels partenaires? *Politique étrangère*, 2, pp. 361–371.

Seeberg, P (2010) *Union for the Mediterranean* – pragmatic multilateralism and the depoliticization of EU–Middle Eastern relations, *Middle East Critique*, 19(3), pp. 287–302.

Tassinari, F. & Holms, U. (2010) Values promotion and security management in Euro-Mediterranean relations: 'making democracy work' or 'good-enough governance'?, DIIS Working, 17.

# The UfM's Institutional Structure: Making Inroads towards 'Co-Ownership'?

ELISABETH JOHANSSON-NOGUÉS
Institut Barcelona d'Estudis Internacionals (IBEI), Spain

ABSTRACT *The Union for the Mediterranean (UfM) has been furnished with a more ambitious institutional structure compared to the Barcelona Process, in the hope that this will provide non-EU partners with a greater say over co-operation processes. This contribution examines the different components of the institutional structure and explores the UfM's potential for achieving its stated ambition to reinforce co-ownership. The essay concludes, nevertheless, that a host of obstacles currently stand in the way of making inroads into co-ownership in Euro-Mediterranean relations.*

As the Union for the Mediterranean (UfM) began to take shape on the drawing board, 'co-ownership' quickly emerged as a distinctive feature of the new initiative. The UfM's predecessor, the Barcelona Process, had regularly been accused of being excessively skewed toward the European Union's structures and preferences. As French President Nicolas Sarkozy and his team looked to revamp and reinvigorate relations across the Mediterranean, they declared the EU's protagonist role since 1995 to have been so strong as to stifle any potential initiative from southern Mediterranean partner countries and therefore, to their mind, one of the main culprits for the Euro-Mediterranean Partnership's lacklustre co-operation record. The UfM architects therefore deemed an institutional recalibration among northern and southern partners necessary.[1] Co-ownership, it was hoped, was going to entice southern partners enough to turn them from 'unhappy laggards' into 'low profile supporters' of Euro-Mediterranean co-operation under its new guise and give relations a new boost (see Bicchi, this collection).

However, what has happened since with this ambition to foment co-ownership? This contribution will look at the UfM institutional structure and explore its potential for living up to the ambition to provide partners with genuine co-ownership. The first section will provide the background and a brief conceptual discussion related to co-ownership in consensus decision-making settings such as the UfM. The second section will supply details of the different bodies making up the UfM institutional structure and the third and final section will attempt to analyse whether the institutional architecture is making inroads towards its objective to instil co-ownership of Euro-Mediterranean co-operation.

**From Barcelona to Paris**

Among the main novelties when the Euro-Mediterranean Partnership was launched in Barcelona in 1995 was the idea of holding periodic institutionalized multilateral meetings.[2] At the top of the political steering of the Barcelona Process were the Euro-Mediterranean Conferences of Ministers of Foreign Affairs which, meeting initially biennially but from 2001 onwards largely biannually, had the task to provide direction and to monitor the implementation of the Partnership's Work Programme (Philippart, 2003). The Euro-Mediterranean Committee for the Barcelona Process (the Euro-Med Committee), composed of senior officials and meeting bimonthly, was charged with the review and follow-up of the agenda of the Partnership. The sectoral and diverse thematic meetings of ministers, senior officials and experts provided the specific direction to various areas of the Barcelona Declaration. The European Commission saw to the preparation of meetings in the areas where it had competency, and in co-ordination with the Council of the EU working groups on topics of mixed competences (Edwards and Philippart, 1997: 475). The Council working groups also prepared meetings in the political and security basket of the Barcelona Process, which until 2000 centred around the attempts to set up the 'Euro-Mediterranean Charter for Peace and Stability', and since 2002 have essentially tackled issues related to police co-operation (Gillespie, 2002; Bicchi, 2006) and civil protection (Bremberg, 2007).

These institutional arrangements would initially be heralded as a new departure in Euro-Mediterranean relations for appearing to give southern Mediterranean countries a chance to become full partners of the co-operation processes through their presence and input at meetings. However, the institutional set-up would eventually come under attack from critics on both sides of the Mediterranean for not being representative enough of southern Mediterranean partner concerns (Schmid, 2002). Since meetings were chaired by either the EU Presidency or the European Commission depending on the subject matter, these institutional bodies in effect controlled the agenda and, to some extent, even the outcome. The southern Mediterranean countries regularly voiced complaints that they were not kept sufficiently or timely enough informed with regard to the agenda of the Euro-Med Committee. The lack, or lateness, of information made it difficult for them to elaborate alternative options to the EU's proposals, had they so desired (Edwards and Philippart, 1997; Monar, 1998).[3] Another common complaint was that the 'joint statement' issued after ministerial conferences was authored by the EU Presidency rather singlehandedly, and frequently, as Aliboni (2009: 1) has it, included 'items which were not truly shared by all members but towards which the house guests showed acquiescence, condescension or complacency'. As a consequence, as Bicchi (2006) has argued, there was a perception that European representation and preferences had come to predominate in Euro-Mediterranean relations.

The EU and its member states were aware of the impression that the institutional structure of the Barcelona Process was slanted in favour of the interests of the European Union. There were several attempts from 2002 onwards to remedy this imbalance and to achieve greater parity. However, they either did not gel or were not deemed far-reaching enough (Gillespie, 2002).[4] Hence, when Sarkozy first launched the idea of a re-configuration of trans-Mediterranean relations in the form of the so-called 'Mediterranean Union' during his 2007 electoral campaign, the issue of institutional power asymmetry between the EU and its partners and its alleged detrimental consequences on co-operation was a central concern (Soler i Lecha, 2008). Co-ownership was therefore presented as the remedy to ailing Euro-Mediterranean co-operation and this line of argument remained a central aspect even as the Mediterranean Union proposals eventually evolved into the UfM in 2008.

What was meant by co-ownership in the context of the refurbished Euro-Mediterranean institutions was left rather vague, however. The constitutive UfM documents and later documentation reveal few clues as to how the concept of co-ownership is to be interpreted. Co-ownership is a term which is perhaps most frequently associated with international assistance and development co-operation, but definitions of its meaning tend to abound. In a generic sense, co-ownership is used in contexts where a partnership is struck between donor and local actors enabling joint ownership over resources or project management. Stated differently, one could infer that co-ownership aspires to empower all involved stakeholders to exercise joint influence either across the board in co-operation processes or in relation to very specific project execution.

In terms of the UfM, the French UfM architects have supported a maximalist reading of co-ownership and close to Sarkozy's early visions of laying 'the foundations of a political, economic and cultural union founded on the principles of *strict equality*' between the EU-27 and its Mediterranean partners (*EUObserver*, 2007; emphasis added), empowering partners in all areas of co-operation. Henri Guaino, head of the French inter-ministerial mission of the UfM, has underlined the necessity 'that power be transferred from the EU and the Commission to the governments in the framework of a body in which southern [partner] countries can make their own decisions, undertake initiatives and if need be, say "no" to European proposals' (Aliboni, 2009: 4). Co-ownership should therefore, according to the French, embrace functional parity on agenda-setting and decision-making for all partners and at all levels of the UfM in the most classic sense of intergovernmental co-operation based on consensus.

Consensus in the intergovernmental setting is generally looked upon as the best guarantee for a sovereigny-sensitive state to safeguard its preferences, given the ease with which even the smallest member state can prevent common action by imposing its veto. Such empowerment over decision making can, according to the Institutionalist school of thought, foment a virtuous circle among co-operating states based on solidarity, trade-offs and/or side-payments, where indeed a perception of co-ownership prospers as many, though not all, EU policies taken in an intergovernmental mode testify (Scharpf, 2006). The key to the success of consensus-based joint decision systems and, by extension, co-ownership, appears to be the presence of trust among co-operating states. In the absence of such trust, the Institutionalist literature warns us, the joint decision system based on consensus can become a downward spiral of decision-making 'traps' and eventually institutional paralysis (Ibid.). Two of the most common problems affecting such intergovernmental co-operation are the existence of veto-players, impeding an upgrade of common preferences, and the weakness in those common institutions which are designed to overcome veto-playing by their 'honest brokerage' capabilities. These problems, both present in the context of the UfM as we shall see, have in turn wide-ranging implications for a sense of co-ownership over co-operation processes.

## The UfM Institutional Structure

After months of French diplomatic footwork to convince partners on both sides of the Mediterranean of the necessity to upgrade relations and co-operation structures, the heads of state and government of all 43 partners finally met in Paris in July 2008. They agreed there on the convenience of establishing 'new institutional structures which will contribute to achieving the political goals of this initiative, especially reinforcing co-ownership, upgrading the political level of EU–Mediterranean relations and achieving visibility through projects' (Paris Declaration, 2008). They also sketched out the contents of the new UfM institutional structure, and called upon the foreign ministerial meeting scheduled to take place in Marseille later in the year to work out all the practical details. This section outlines the different institutional bodies as they have developed since (see Figure 1).

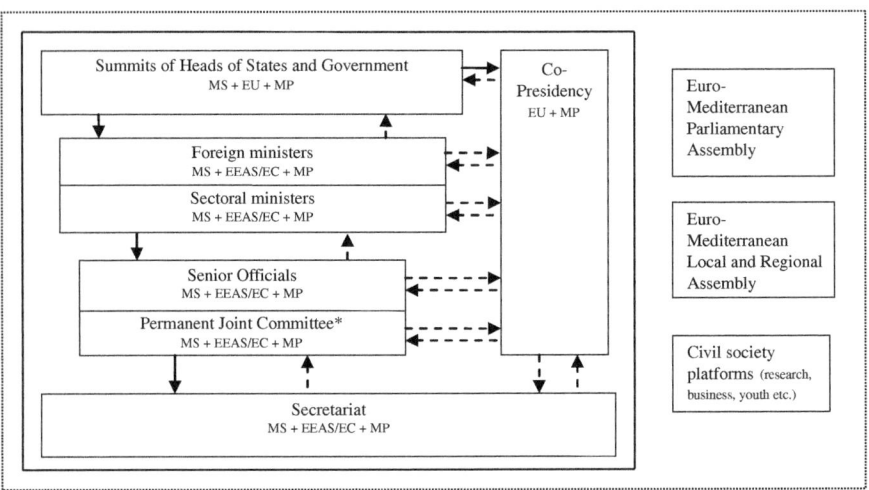

**Figure 1.** The UfM institutional structure.

## Summits of Heads of State and Government

The inaugural UfM Paris summit in 2008 launched the practice of high-level meetings among the heads of state and governments from the 43 partner countries and relevant representatives from the European institutions. The heads of state and government and EU representatives are to meet biennially with the objective of providing political guidance for the UfM and establishing a two-year work programme consisting of prioritized areas of co-operation.

The summits represent a novelty compared to the Barcelona Process. The heads of state and government have now assumed the formal political steering of the initiative, a role, as we have seen, held by the foreign ministers in Euro-Mediterranean Partnership. The decision to meet at the highest level of state and government stems, in part, from a desire to give the EU's Mediterranean partners the same political importance accorded to Asian or Latin American countries, with which the Union regularly holds summits (Aliboni *et al.*, 2008: 23). In part, the decision to hold summits is also a reflection of the fact that political initiative in many partner countries lies very much in the hands of the heads of state and government and hence having their explicit, active and even personal involvement may facilitate co-operation by exacting a commitment to supporting the implementation from highest level of state authority.[5]

However, the benefits of having the heads of state and government involved for the higher political impetus they may bring to the UfM table must at all times be contrasted with the possibility for even further politicization of Euro-Mediterranean relations (Gillespie, 2008: 281). The presence of heads of state and government is much more emblematic of the state of relations among countries, compared to foreign ministers, and is usually the first thing to be cancelled in situations of inter-state tension, as occurred twice in 2010 with the June and November summits scheduled to be held in Barcelona. The continued conflictive relations between many of the non-EU UfM partners therefore give grounds to think that the formal inclusion of heads of state and governments in an institutionalized setting could make the UfM even more vulnerable to the shifting political winds of the non-EU UfM partners compared to the Euro-Mediterranean Partnership.

## Co-Presidency

One of the other main institutional novelties launched by the Paris summit is the UfM co-Presidency. The co-Presidency, composed of one EU co-president and one southern partner co-president, is charged with the task of calling and chairing all summits, all ministerial meetings, senior officials' meetings and, when possible, other experts or ad hoc meetings within the initiative. The co-presidencies will submit the agenda to all parties for approval and work to obtain consensus for the common conclusions of summit, ministerial, and other meetings. France and Egypt assumed the UfM co-Presidency in 2008.

The co-Presidency ends the practice of the EU Presidency and Commission presiding over Euro-Mediterranean meetings as they did in the Barcelona Process.

Formally, the co-Presidency also offers partner countries the potential for more leverage over agenda-setting and the negotiated outcome. The intention of the co-Presidency is that it may function similarly to a parliamentarian 'whip' to achieve and/or accelerate agreement on issues, by being continuously updated by partners and by performing the task of informing, cajoling and arm-twisting in just the right proportions. In theory therefore, the role of the co-president has, if exploited deftly, all the trappings of enabling EU and southern UfM partners alike to put a decisive imprint on the co-operation scheme and have a say over the proposals put forward within the UfM framework.

However, it is worth noting that, at the time of writing, the full details of the functioning of the EU component of the co-Presidency have yet to be revealed. According to the Marseille Declaration (2008: 4), the EU part of the co-Presidency 'must be compatible with the external representation of the European Union in accordance with the Treaty provisions in force'. In other words, in 2008 this consisted of the rotating EU Presidency and the European Commission (under the Treaty of Nice) and, with the Lisbon Treaty, the EU element of the co-Presidency should consist of the troika composed by the President of the Council of the EU, the High Representative of the Union for Foreign Affairs and Security Policy and Vice-President of the European Commission (HRVP), and the rotating EU Presidency. This stipulation notwithstanding, what has transpired since the Paris and Marseille Declarations is a sort of EU 'co-co-presidency', with France acting as 'senior' co-president, in lieu of, or together with, the rotating EU Presidency. Balfour (2009: 102) has argued that this came about because the Paris and Marseille Declarations did not reflect French preference on this point. Prior to the 2008 Paris summit France had tirelessly advocated that the period of tenure of the European part of the co-Presidency should match the two-year stint of the non-EU part. It is reported that the French therefore took the initiative to negotiate a bilateral deal with each incoming EU Presidency to be able to hold on to the co-Presidency until end of the two-year term. The co-co-presidency practice, which has caused some controversy among EU member states for not adhering to Treaty provisions, could end with the next summit of heads of state and government. Although no firm decision has yet been reached, there are indications that the HRVP and the European External Action Service (EEAS) may nominally assume the 'northern co-presidency' in 2010–12 in the name of the troika. The HRVP will in turn choose a deputy to perform this function. This deputy is currently yet to be confirmed, but there are some suggestions to the effect that this task may fall upon Spain (Congreso, 2010: 27).

The co-Presidency is a clear example of institutional change compared to the Barcelona Process, with meetings now being prepared and chaired by representatives from both sides of the Mediterranean. The creation of the figure of the co-Presidency therefore appears, at least on paper, to have redressed the EU's erstwhile predominance in Euro-Mediterranean relations. However, and as we will explore further in the analysis, the way that the office of the co-Presidency has evolved since 2008 has contributed to more turbulence than clarity in terms of how it can help bring about a truly recalibrated institutional landscape among partners, beyond the mere formality of its creation.

## THE UNION FOR THE MEDITERRANEAN

*Ministerial Meetings, Senior Officials and Joint Permanent Committee*

The Paris Declaration holds that '[a]nnual Foreign Affairs Ministerial meetings will review progress in the implementation of the summit conclusions and prepare the next summit meetings and, if necessary, approve new projects'. This marks continuity with the Barcelona Process, although their formal weight in the steering of the UfM has been reduced as a consequence of the involvement of heads of state and government. The ministers now only adopt the necessary guidelines in accordance with the mandate established by the work programme. Sectoral ministerial meetings will also continue in the same fashion as during the Euro-Mediterranean Partnership. However, the frequency of their meetings has been reduced, at present scheduled to meet annually instead of the Euro-Mediterranean Partnership's biannual meetings (regular meeting plus mid-term meetings). The sectoral ministerial meetings have, however, arguably become perceptibly more important to the UfM compared to what they were under the Barcelona Process formula. This is in part due to the main focus of the UfM being, above all, issue-specific thematic areas (e.g. the maritime de-pollution, maritime and land highways, or small and medium enterprises). The sectoral ministerial meetings could gain even more relevance if and when the work of the UfM Secretariat to spearhead these co-operation initiatives begins in earnest. In part the importance attached to sectoral ministerial meetings is also because when other meetings have been cancelled due to the Middle East tension, several sectoral meetings have still taken place (e.g. water, sustainable development projects, economy and finance, women, trade, employment and health; cf. Martín, 2010).

In terms of the senior officials, the Paris Declaration envisions that they continue their task to prepare the ministerial meetings, project proposals and the annual work programme as under the Barcelona Process, albeit now with an expanded mandate to deal with the full range of aspects related to the UfM. In effect, with the UfM the senior officials have acquired a more prominent role in the institutional structure and in the political steering of the co-operation processes for two reasons. First, there is no longer any formal separation between political and security issues on one side and socio-economic matters on the other, as under the Barcelona Process. The senior officials will now take stock of and evaluate the progress of the UfM in all its components, including the socio-economic issues previously handled by the Euro-Med Committee. Second, the regularly held senior official's meetings have become the main means by which to advance the political aspects of the UfM co-operation processes since 2008 as a consequence of the semi-paralysis of other UfM institutional bodies and the suspension of foreign ministerial meetings.

As for the Joint Permanent Committee (JPC), the Paris Declaration establishes that the committee, composed of national representatives based in Brussels, is to assist and prepare the meetings of the senior officials and ensure the appropriate follow-up. The JPC may also act as a rapid reaction cell if an exceptional situation arises in the region that requires urgent consultation among Euro-Mediterranean partners. The committee is an institutional novelty compared to the Barcelona Process although it is not yet clear if and when it will be convened for the first time and how exactly it will function.

Most of these institutional bodies and working parties mark the continuation of the Barcelona Process, albeit some with a modified mandate. The perhaps most visible difference is that now their meetings are chaired by the co-Presidency. Another important difference is that, in the spirit of institutional symmetry between the EU and its partners, no concerted EU official position is adopted prior to UfM ministerial or lower level meetings. This ends the practice under the Barcelona Process that the EU member states and the European Commission met before each Euro-Mediterranean meeting to establish the EU's common formal policy position on all issues to be discussed with partner countries.[6]

*Secretariat*

The Secretariat, launched in July 2010 and based in Barcelona, is to work on identifying concrete projects that can be launched in the framework of the UfM, as well as to find the money to sponsor such projects. The Secretariat's statutes (2010) hold that the body is to function 'in operational liaison with all structures of the process, particularly with the co-presidencies, including by preparing working documents for the Senior Officials, and through them for the other decision making bodies'. The Secretariat is to have a separate legal personality with an autonomous status. Its mandate is of a technical nature, while the political mandate related to all aspects of the initiative remains the responsibility of the ministers of foreign affairs and senior officials.

The Secretariat is an institutional novelty compared to the Barcelona Process. The nomination of Ahmed Masadeh, a Jordanian diplomat, as the first secretary-general of the Secretariat was confirmed by the foreign ministers of Egypt, France, Jordan, Spain and Tunisia in January 2010.[7] The Secretariat will also count on the presence of six deputy secretary-generals, each with a portfolio corresponding to one of the six priority issues identified by the Paris Declaration: de-pollution of the Mediterranean, maritime and land highways, civil protection, the development of alternative energies (especially solar energy), higher education, research and the Euro-Mediterranean University, and the Mediterranean business development initiative. Moreover, the work of the secretary-general and the deputies will be assisted by seconded officials in an advisory capacity from relevant national ministries and from the EEAS.

The Secretariat, for all its pretension of being a purely technical – and hence apolitical – body, has nonetheless become first a victim and later the focus of intense political and strategic bargaining. Several ministerial meetings intended to concretize the specific institutional parameters of the Secretariat were scheduled for early 2009 and again for late 2009, but were postponed as a consequence of Arab–Israeli disputes. The Secretariat's start-up date, originally planned for early 2009, was thus inevitable delayed. Moreover, while the nomination of the secretary-general of the Secretariat was easy – Masadeh's was the only candidate – the matter of the flanking deputy secretary-general of the UfM has been highly controversial. The Secretariat was initially planned to have one deputy secretary-general, but by the foreign minister meeting in Marseille in 2008 the proposals had swollen to five: Israel, the Palestinian National Authority, Greece, Italy, and Malta. The Israeli deputy was a quid pro quo for Israel dropping its opposition to the Arab League's request to participate in the UfM as observers (*Libération*, 2008). The Turkish request for a deputy became another controversy later on as Cyprus opposed it for reasons related to the difficult relations between these two countries. The two countries' opposing views on the matter made it impossible for the February 2010 meeting of senior officials to make a decision on the UfM Secretariat's statutes, although the matter was eventually settled at a later senior officials' meeting.

Beyond such polemics, the unclear financial situation of the Secretariat has been another conditioning factor impeding its rapid and successful consolidation. The original blueprints for the Secretariat included a proposal for a permanent annual budget of €10bn provided directly by partner countries. However, so far money has only trickled in to the Secretariat. The economic crisis has hit hard and in times of financial austerity the pledges of UfM partner countries to the co-operation projects have not always materialized, not even France has fulfilled its original pledges.

The Secretariat, intended as an institution that would serve all UfM partners, initially generated expectations that co-operation in the Mediterranean will begin to take on a very different dynamic compared to under the Barcelona Process. The European Parliament (2010: 4) expressed its firm conviction that 'the UfM's secretariat needs to become the [UfM institutional] structure's linchpin. It has the ability to become an autonomous actor and to provide real added value to co-operation across the Mediterranean'. Theoretically, the whole project cycle for different co-operation initiatives launched within the framework of the six priority areas could henceforth be managed by the Secretariat, in representation of the 43 partners, and break with the erstwhile domination of the European Commission. Nevertheless, the limitations imposed on the Secretariat still make it difficult to envision how it could boost co-operation among Euro-Mediterranean partners. To prove itself to be an effective successor to the European Commission, the UfM Secretariat will have to establish itself fairly quickly as the central point of reference for co-operation in the Mediterranean area. However, with vague statutes, no firm budget and with highly complex thematic briefs requiring expertise and manpower much beyond the limited human resources currently pledged to the Barcelona-based body, it is still unclear how the Secretariat can acquire such centrality.

*Other UfM Institutional Entities*

The Paris Declaration acknowledges the role that parliamentarian, local and non-governmental actors have played in the Barcelona Process. The heads of state and governments 'underscore the importance of the active participation of civil society, local and regional authorities and the private sector in the implementation of the [UfM]' and 'strongly support' the strengthening of the parliamentary dimension of the same (Paris Declaration, 2008). The statement from the Paris summit even goes so far as to state that '[t]he ultimate success of the initiative also rests in the hands of citizens, civil society and the active involvement of the private sector'. The Marseille Declaration (2008) reiterates this support, by affirming that '[c]ivil society should be further empowered and its capability enhanced through improved interaction with governments and parliaments'. Co-ownership of co-operation processes across the Mediterranean is thus not only sought at the intergovernmental level, but also through other actors involved.

However, more points to continuity with the Barcelona Process than a genuine breakthrough. The 1995 Barcelona Declaration also emphasized the importance of civil society and parliamentarian co-operation, but concrete proposals for how these non-state actors could be drawn closer to Euro-Mediterranean institutional frameworks were never forthcoming. The UfM has not produced a clear road map for these actors in terms of how their work and proposals can be better reflected in UfM intergovernmental decision making and hence gain centrality in Euro-Mediterranean co-operation (Balfour, 2009: 103).

The parliamentary and non-governmental actors have, however, continued to develop outside the formal UfM institutional structures as loosely affiliated bodies (see Figure 1). In terms of inter-parliamentarian co-operation, the Euro-Mediterranean Parliamentary Assembly (EMPA) is trying to expand its current focus on mostly sectoral issues and adopt a more political dimension. Moreover, the EMPA and the Parliamentary Assembly of the Mediterranean (PAM) are trying to find ways of collaboration. The EMPA draws parliamentarians from the 43 states of the UfM and the European Parliament while the PAM is more limited in scope, only including Mediterranean rim states. Here the main bone of contention seems to be that some EU member states would like to see a strengthening of the EMPA, while others such as France prefer giving support to the PAM.

In terms of regional and local actors, they have met occasionally in the framework of the Barcelona Process, but not under formal circumstances. However, in 2010 the Euro-Mediterranean Regional and Local Assembly (ARLEM) was inaugurated in Barcelona. ARLEM's institutional framework consists of two co-presidents: Luc Van den Brande, the EU's Committee of the Region's President, and Mohamed Boudra, mayor of the Moroccan city of Al Hoceima. The Assembly will count on the technical support of a Barcelona-based secretariat, housed in the same premises as the UfM Secretariat. The 2010 work programme which was adopted at the inaugural meeting focused on urban and territorial development, decentralization, the information society, support for small and medium enterprises, local water management, migration and cultural co-operation. While maintaining its institutional autonomy, however, ARLEM aspires to link up with the other UfM institutional bodies by ways of observer status. Such a status would allow representation at relevant UfM meetings and provide opportunity for expressing policy positions in such fora.

Finally, it has been frequently noted that the single, lasting success of the Barcelona Process was the many civil society networks it inherited from the Renewed Mediterranean Policy and later boosted during the decade which followed the adoption of the Barcelona Declaration. The Paris and Marseille Declarations commend the work carried out by civil society networks. However, the two Declarations do not heed the longstanding demand from civil society actors that they be consulted prior to ministerial meetings. The UfM therefore follows the path already staked out by the Barcelona Process whereby civil society co-operation is encouraged, but provides no firm decision on how to derive concrete synergies from their work to assist the development of specific ambits related to the UfM.

The UfM therefore exhibits many of the same characteristics as the Barcelona Process in terms of its relation with parliaments, local governments and non-state actors. There is a continued (virtually hermetic) separation between the intergovernmental dimension and other actors involved in Euro-Mediterranean co-operation, which is essentially a tribute to governments' unwillingness to be too constrained on foreign policy choices by parliamentary scrutiny or held up by lengthy deliberative processes with lower levels of government. Moreover, there are no signs of any advances in political will among southern governments to lend support to civil society organizations today compared to under the framework of the Barcelona Process. Non-democratic regimes on the southern and eastern side of the Mediterranean continue to be reluctant to share the political space with associations which might over time become contestants for political power (see Schlumberger, this collection; Johansson-Nogués, 2006).

## Making Inroads towards 'Co-ownership'?

The above survey of the different institutional bodies of the UfM reveals a highly fragmented institutional landscape. There are still many parameters which are unknown in terms of how the different components of the UfM institutions will function individually and/or together. This makes it much more difficult to assess whether the new UfM institutional set-up has contributed to reinforcing co-ownership and what effect, if any, this has had on co-operative practices.

There are numerous factors which account for this state of affairs. However, one of the principal reasons has of course been the deteriorating situation in the Middle East in the period since the 2008 Paris summit. The Israeli incursion into Gaza in 2008 led to some partners deciding to stay away from the planned UfM meetings of ministerial and senior officials in early 2009, in effect bringing the UfM to a standstill. Some sectoral meetings were resumed in summer 2009 and senior officials' meetings later on, but they continue to be conditioned by the situation in the Israeli–Palestinian conflict. No UfM foreign ministers' meeting has been held since Marseille in 2008. The Middle East tension has thus become an important obstacle to consolidating the UfM's institutional structure, given that summits and the many lower levels needed meetings to flesh out the details of the functioning and mandate of the different UfM bodies simply have not taken place. The unconsolidated state of the UfM's institutional set-up, in turn, impedes the ambition to make bold advances in terms of co-ownership of co-operation processes.

However, it is fair to say that the ambition for fomenting co-ownership in Euro-Mediterranean co-operation also suffers from problems derived from other bottlenecks apart from the Middle East conflict. One such bottleneck is the sclerosis which the UfM exhibits as a consequence of problems stemming from its classic intergovernmental consensus-based co-operation, such as, for example, veto-players and weaknesses inherent to the two institutional bodies which were specifically created to reinforce co-ownership. Hence, in spite of the UfM's ambitious remit to move beyond status quo and the Barcelona Process in terms of co-ownership it finds itself facing an uphill battle to do so.

First, in terms of veto-playing it has affected virtually all levels of the UfM. The most blatant expression of veto-playing has been the different 'empty-chair' crises since 2008, whereby one or several partners have impeded the holding of summits and ministerial or other meetings as a protest over events in the Middle East, as noted above. Other meetings have been held up for other reasons, Israel started the trend by opposing the holding of meetings between July and October 2008 to show its disagreement with the proposal to grant observer status to the Arab League and, in October 2008, Jordan retaliated by postponing an important Euro-Med conference on water security to pressure the other UfM member states to accept the Arab League's inclusion as observer. Intergovernmental veto-playing also keeps associated UfM institutions, such as the Euro-Mediterranean Parliamentary Assembly, ARLEM and the various civil society groupings, at an arms-length from the central decision-making processes of the UfM. Intergovernmental concerns over being constrained by empowering these actors too much is a clear impediment for these actors as they seek share co-ownership of the UfM with governments in the short to medium term. However, veto-playing has not only affected the grand institutional design of the UfM, but also in the minutiae of day-to-day work. An illustrative example of this was the sectoral ministerial meeting on water in April 2010, where the participants agreed on all the technical aspects in terms of co-operation on water, but the agreement fell on the concrete wording in terms of referring to the Palestinian occupied territories. In other words, the presence of veto-players contributes to a situation where all partners see their theoretical ability to have a say and co-ownership of the UfM processes as highly contingent on the highs or lows of the regional political barometer. Hence, one could argue that their co-ownership is, as a consequence of the decision making by consensus, more intermittent and ad hoc than a permanent and consistently applied principle. While the possibility to explore 'variable geometry' and constructive abstention figured in some of the early UfM documents, especially in reference to specific projects, in reality this has been much more difficult to achieve than initially foreseen. Several southern partners (and even some EU members) fear that abandoning unanimity, even on minor projects, would entail an end to their strategic use of veto against initiatives, whether current or future, which are deemed not to be in line with their particular national interest and/or might favour a regional rival. Syria, for example, has been one of the more conservative in terms of any initiatives to side-step the unanimity rule, due to its wish to be able to micro-manage policy evolution on the Middle East conflict and relations with Israel.

Second, in terms of the UfM institutions, the two institutional bodies – the co-Presidency and the Secretariat – which were specifically intended to give co-operation and co-ownership a boost, have been riddled by a set of confusing or weak mandates. Their ability to carry out the role of 'honest brokers' and to prod co-operation forward has thus been impaired. As for the co-Presidency, in comprising one EU and one southern partner representative, the very creation of this institutional body as of the Paris summit seemed to herald a leap forward toward greater co-ownership. However, any expectations of co-ownership beyond this mere formality – for example, that the French–Egyptian duo would endeavour to devise a system where most, if not all, stakeholders felt themselves to be participants and empowered – have so far been frustrated. Some of the co-Presidency's practices rather point in the opposite direction. As for the northern co-Presidency, more than one EU member has come to perceive UfM as excessively a 'French project' rather than a shared one (Aliboni, 2009). In particular, active French use of the EU element of the co-Presidency to push the UfM down the road of its own intergovernmental reading has troubled some EU members as they still consider the UfM essentially an EU policy and therefore think it should conform to Treaty stipulations. This is evident from the controversies which have arisen around whether the EU should or should not meet prior to UfM meetings, however informally (to which France reluctantly had to agree to in the end; see note 6), and the permanence of France in the role of the northern co-president (where the French vision has imposed itself). Sidestepping the EU Lisbon Treaty is a matter which has especially worried Germany, Berlin's view being that 'it is necessary that – within the Union for the Mediterranean – the EEAS will speak for the European countries' because '[w]e do not want southern European countries to focus exclusively on the southern border and eastern European countries doing the same on the eastern border. This would lead to a division of the European Union' (Hoyer, 2010).

There are therefore indications that some EU partners do not feel fully and adequately represented in the UfM by their French co-president. Their notion of the UfM sits uneasily with the policy line pursued by Paris and this fact contributes to their perception of loss of ownership over Euro-Mediterranean co-operation processes when compared to the Barcelona Process. For that reason, should the current proposal to have the upcoming EU co-president nominated by the HRVP finally prosper, and if that co-president chooses to collaborate closely with the EEAS, such perceptions of marginalization and sub-optimal representation might be substantially defused. However, it is fair to say that until a definition of the co-Presidency office is definitively settled such unease will continue among EU member states.

There is a similar situation in terms of the questionable representativeness of the southern co-president. The designation of Egypt as the first southern co-president was not only important to get that country on board and in favour of the initiative: Cairo was also 'keen to confirm its political leadership in the Middle East' and saw in UfM a means to enhance its regional leader credentials (Balfour, 2009: 102; Comelli, 2010). However, if the initial ambition was to ensure mutually supportive linkage between the UfM and the Middle East, the Egyptian government has since the Paris summit steered a different course. With the Israeli incursion into Gaza in 2008, Cairo announced, at other Arab states' behest, the suspension of all UfM activities in protest against the Israeli military action (Sénat, 2009). Egypt's threat of a boycott was later one of the factors behind the decision to postpone several other meetings in 2009 and 2010, for example the 2009 foreign ministerial conference and the 2010 Barcelona summit, as a consequence of some Arab countries' reluctance to meet up with Israeli foreign minister Avigdor Lieberman (*AFP*, 2010) and the failure of Middle East peace talks. Thus, in spite of its role as a co-president and its vow to uphold interests of all members of the UfM, Cairo has acted more as a champion of the Egyptian national interest and that of a small group of Arab UfM members (Balfour, 2009; Khatib, 2010). Turkey, in particular, has voiced complaints that Cairo has only consulted with and acted for the interest of some of the Arab UfM members, as opposed to the benefit of all southern partners (Sénat, 2009). However, it is also true that the southern partners are not an easy group to represent. The heterogeneity of the southern partners' social and economic situations is not very conducive to preference coalitions in the UfM co-operation processes. The differentiated socio-economic realities (e.g. Turkey/western Balkans vs. the Arab partners) or inter-state tension (e.g. Middle East, Western Sahara) are thus conditioning factors speaking against the non-EU partners developing and actively pursuing a solid co-ownership of the UfM through the southern co-Presidency in the near to medium future. The non-Arab UfM partners' lack of organizational ties outside the UfM is an additional shortcoming, since they have no common forum in which to negotiate a harmonized position with which to influence the southern co-Presidency in one way or the other. This disparity of views has even hampered the search for a new southern co-president to replace Egypt for the upcoming non-renewable period of two years (2010–12) as the Paris and Marseilles Declarations stipulate. All indications therefore appear to point to the more than probable continuity of Egypt as the southern co-president even for the coming term for want of alternative candidacies. It is thus left to be seen whether the adequate representation of non-Arab and non-Mashreq UfM partners will be improved upon in Cairo's upcoming second term co-Presidency in order to better guarantee the voice opportunity of all UfM southern partners and improve their perception of ownership.

Another example of institutional weaknesses which raises queries about the UfM's ambition to reinforce co-ownership is the Secretariat. Launched only in 2010, even then it did not immediately become functional. Indeed, it continues to have unresolved questions as to its function and how it can become the expected 'linchpin' of Euro-Mediterranean co-operation and ensure a distinct improvement over the Barcelona Process. It is thus unclear what kind of institutional body the UfM partners now theoretically have co-ownership over and whether such ownership will be worthwhile over time. The size of the Secretariat's budget and the use it will make of it are probably the two factors that will provide the clearest litmus-test of the future added value of this institution. However, if southern partners truly aspire to make inroads in terms of co-ownership, their individual contribution to the Secretariat's global fund sustaining the six UfM priority projects must be proportionally substantial to be able to assume a steering position over these projects. The trend set by some projects flagged under the UfM umbrella, like the Mediterranean Solar Plan, is that they appear to continue along the paths staked by the Barcelona Process whereby co-operation is dominated both technically and financially by European business and large public donors (European Commission) to the extent that local, non-EU, initiative and/or decision-making powers are scarce if not non-existent. The lack of participation in terms of providing for adequate and balanced financing is thus a major obstacle for realizing the ambition of co-ownership and making a qualitative leap forward in Euro-Mediterranean relations.

Another and final bottleneck which speaks against a quick fix on empowering all UfM stakeholders is the waning interest in the principle of co-ownership that can be detected among non-EU UfM partners. It is fair to say that the prospect of co-ownership over multilateral Euro-Mediterranean structures for a majority of non-EU UfM partners no longer holds the same appeal as it used to under the early years of the Barcelona Process. The reason for this waning interest is twofold. First, bilateral relations have come to take precedence over multilateral for some partner countries (Gillespie, 2008: 285). The launch of the European Neighbourhood Policy (ENP) was heavily criticized for its EU-instigated reforms (Johansson-Nogués, 2004). However, the pragmatism which has ensued in the practical consultation for the ENP Action Plans and later its implementation has meant that non-EU partners today perceive themselves as having a fair amount of ownership of their respective bilateral relations with the EU (Driss, 2009: 5; Khatib, 2010). The UfM's multilateral format and consensus rule, even with the co-ownership principle, cannot provide the same perception of parity and control experienced at the bilateral level and hence detracts from its value in non-EU partners' eyes.[8] What is more, for some Arab UfM partner countries bilateral relations are preferred given that they do not oblige them, as the UfM's multilateral format does, to share decisions, with Tel Aviv or any other regional rival, in terms of what projects should be implemented, and how. Finally, non-EU partners have in recent years developed and/or strengthened alternative bilateral relations with Asian and, above all, Gulf countries which serve to reduce their need and/or interest for engaging with the EU and the multilateral UfM (Schumacher, 2010). The relevance of co-ownership and multilateral political frameworks in current Euro-Mediterranean relations thus might have been surpassed by a set of dynamics of fairly recent history.[9]

**Conclusions**

The launch of the UfM was supposed to mark a before and after in Euro-Mediterranean relations. It was hoped, especially in the EU, that the new institutional structure would boost co-operation with the southern partners and motivate a greater commitment from non-EU partners to UfM projects. The UfM especially sought to correct the perceived EU bias in the Barcelona Process by introducing the notion of co-ownership. If the Barcelona Process was regularly accused of being an EU-centric exercise, the UfM promised to make a clean break with the past by introducing the notion of greater parity among all participant states.

However, it has so far proven to be a complex task to make inroads towards the principle of co-ownership and to provide a decisive added value over the Barcelona Process co-operation. The Middle East tension has been the most obvious obstacle to rapid progress. However, there are other important bottlenecks hindering the further development of the UfM and the objective of creating more co-ownership for all UfM partners. Veto-playing has been rampant and the implementation of the institutional structure since the Paris summit has so far failed to resolve some significant confusion among partners about several key institutional bodies' function and purpose. The much vaunted greater voice opportunity for non-EU partners over Euro-Mediterranean decision-making processes has therefore in many instances yet to consolidate into a single, standardized and workable practice. Another important impediment in the way of co-ownership is the current lack of interest shown for this principle by many non-EU UfM partners. The principle of co-ownership has thus not proven to be a sufficiently big 'carrot' to entice southern partners to shift from their erstwhile unhappy laggardness into the 'low profile supporters' of Euro-Mediterranean co-operation and stimulate accelerated co-operation processes as the French UfM architects had initially hoped.

The fragmentation and drift which characterized Euro-Mediterranean co-operation under the final years of the Barcelona Process therefore appear to continue even under the revamped UfM institutional structure. On the European Union's side there is some expectation that the EEAS, once fully operative, will help to resolve some of the current Gordian knots that exist among EU member states in terms of the UfM. On the non-EU UfM partner side it is, however, more difficult to foresee solutions in the short to medium term for the great heterogeneity which characterizes their socio-economic situation and for the lack of common fora for harmonizing policy positions. Should there be a way to overcome these structural problems, there are still more questions to be resolved before the UfM can begin to make inroads on co-ownership. Is it possible that the UfM institutional structure can be said to have given too much premium to formal ownership (legitimacy), over actual ownership and concrete policy output (efficacy)? Could the re-engagement of southern partners to the UfM and a multilateral co-operation institutional structure therefore paradoxically be found in a more adequate balance between greater legitimacy (co-ownership) and output (tangible co-operation projects, perhaps even some form of 're-Europeanization' of the initiative)? However, such recalibration appears at the moment not to form part of the debates inside the meeting halls of the new UfM institutions on the few occasions they meet.

## Acknowledgements

The author is a member of the Observatory of European Foreign Policy, Barcelona, Spain (http://www.iuee.eu) and wishes to acknowledge the financial support of the *Programa Nacional de Movilidad de Recursos Humanos de Investigación*, Ministerio de Ciencia e Innovación, Spain.

## Notes

[1] Northern partners here refers to the EU-27 and the southern partners encompass: Albania, Algeria, Bosnia and Herzegovina, Croatia, Egypt, Israel, Jordan, Lebanon, Mauritania, Monaco, Montenegro, Morocco, Occupied Palestinian Territories, Syria, Tunisia, and Turkey.

[2] The Euro-Arab Dialogue which functioned between 1973 and 1979 was on the EC side only represented by the EU troika. The Renewed Mediterranean Policy launched in 1990 did not contemplate any multilateral meeting component (Gomez, 2003).

[3] Monar (1998: 55) explains the Commission's lateness as being a consequence of a chronic under-staffing and excessive workload of the DGs responsible for agenda.

[4] A form of co-chairmanship of the Euro-Mediterranean Committee was proposed at the Valencia ministerial meeting in 2002, but in the final communiqué there was only a commitment to studying the idea. Aliboni (2008: 9) also points to the circulation of several non-papers among senior officials in 2006 and 2007 containing provisions to make the Barcelona Process more efficient by instituting a North–South Presidency and a Secretariat.

[5] Philippart (2003), writing against the backdrop of then existing academic proposals to hold summits of heads of state and government, considers that '[t]he direct involvement of the top players would undeniably lower the risk of organised non-compliance. As long as the EMP remains restricted at ministerial level, it will be relatively easy [for the ministers] to play the clock or come back on any agreement, using the pretext of a mandate breach'.

[6] Informal EU meetings used for consultation and for sounding out positions among EU member states, do, however, take place.

[7] Given the freeze on UfM foreign ministerial meetings as a consequence of the Gaza invasion, this ad hoc meeting took place in order to try to unblock the UfM and move forward on a project (the Secretariat) which at the time most UfM partners appear to favour. Masadeh resigned in January 2011.

[8] There are, on the contrary, some fears that the UfM's multilateral component could have a detrimental impact on bilateral relations between the partner country and the EU. This is well-illustrated by the coolness with which the King of Morocco greeted the original proposals of the UfM (Emerson, 2008), fearing that it would upset Morocco's ambition for an advanced status (finally signed in October 2008, see Martín, 2009).

[9] For a more detailed account to the complexities for EU-supported multilateral co-operation as a consequence of bilateral considerations, see Johansson-Nogués (2009).

## References

AFP (2010) Mediterranean summit delayed to allow Mideast dialogue, 20 May.

Aliboni, R. (2008) Southern European perspectives, in: R. Aliboni, A. Driss, T. Schumacher & A. Tovias (Eds) Putting the Mediterranean in Perspective, *EuroMeSCo Papers*, 68, June.

Aliboni, R. (2009) The Union for the Mediterranean: evolution and prospects, *Documenti, IAI* 0939E. Istituto Affari Internazionale, Rome, Italy, December.

Aliboni, R., Joffe, G., Lannon, E., Mahjoub, A., Saaf, A. & de Vasconcelos, Á. (2008) Union pour la Méditerranée: Le potentiel de l'acquis de Barcelone, *ISS Report*, European Institute of Security Studies, November.

Balfour, R. (2009) The transformation of the Union for the Mediterranean, *Mediterranean Politics*, 14(1), pp. 99–105.

Bicchi, F. (2006) 'Our size fits all': normative power Europe and the Mediterranean, *Journal of European Public Policy*, 13(2), March, pp. 286–303.

Bremberg, N. (2007) Between a rock and a hard place: Euro-Mediterranean security revisited, *Mediterranean Politics*, 12(1), pp. 1–16.

Comelli, M. (2010) Dynamics and evolution of the EU–Egypt relationship within the ENP framework, *Documenti IAI*, 10/02, Istituto Affari Internazionali, February.

Congreso de los Diputados (2010) *Diario De Sesiones*, Comisiones, IX Legislatura Núm. 611, Madrid, 30 September.

Driss, A. (2008) North-African perspectives, in: R. Aliboni, A. Driss, T. Schumacher and A. Tovias (Eds), Putting the Mediterranean in Perspective, *EuroMeSCo Papers*, 68, June.

Driss, A. (2009) Southern perceptions about the Union for the Mediterranean, *EuroMeSCo Other Research Papers*, June, pp. 1–7.

Edwards, G. & Philippart, E. (1997) The Euro-Mediterranean Partnership: fragmentation and reconstruction, *European Foreign Affairs Review*, 2, pp. 465–489.

Emerson, M. (2008) Making sense of Sarkozy's Union for the Mediterranean, *CEPS Policy Brief*, 155, March.

*EUObserver* (2007) France muddies waters with 'Mediterranean Union' idea, 25 October.

European Parliament (2010) *Draft Report on the Union for the Mediterranean* 2009/2215(INI), Committee on Foreign Affairs, Rapporteur: Vincent Peillon, 26 February.
Gillespie, R. (2002) The Valencia Conference: reinvigorating the Barcelona Process? *Mediterranean Politics*, 7(2, January), pp. 105–114.
Gillespie, R. (2008) A 'Union for the Mediterranean' ... or for the EU? *Mediterranean Politics*, 13(2), pp. 277–286.
Gomez, R. (2003) *Negotiating the Euro-Mediterranean Partnership: Strategic Action in EU Foreign Policy?* (London: Ashgate).
Hoyer, W. (2010) Speech delivered at *European Union – one year treaty of Lisbon*, University of Malta, Valletta, 2 November. Available at http://www.auswaertiges-amt.de/diplo/en/Infoservice/Presse/Reden/2010/101102-Hoyer-Malta.html (accessed 15 November 2010).
Johansson-Nogués, E. (2004) Profiles: a 'ring of friends'? The implications of the European Neighbourhood Policy for the Mediterranean, *Mediterranean Politics*, 9(2, Summer), pp. 240–247.
Johansson-Nogués, E. (2006) Civil society in Euro-Mediterranean relations: what success of EU's normative promotion? *RSCAS Working Paper*, 2006/40, Florence, Robert Schumann Center for Advanced Studies.
Johansson-Nogués, E. (2009) *Post-Power EU? Power and Post-power Practices in the European Union's Relations with Partner Countries in the Mediterranean, Northern Europe and Western Balkans* (Berlin: VDM Publishing).
Khatib, K. (2010) The Union for the Mediterranean: views from the Southern Shores, *International Spectator*, 45(3), pp. 41–50.
*Libération* (2008) Union pour la Méditerranée: Marseille grillée par Barcelone, 4 November.
Marseille Declaration (2008) *Final Statement*, Marseille, 3–4 November.
Martín, I. (2009) EU-Morocco relations: how advanced is the 'advanced status'? *Mediterranean Politics*, 14(2, July), pp. 239–245.
Martín, I. (2010) The priorities of Spain's EU Presidency in the Mediterranean: ideal and reality, *ARI* 34/2010, Real Instituto Elcano, Madrid.
Monar, J. (1998) Institutional constraints of the European Union's Mediterranean Policy, *Mediterranean Politics*, 3(3, Autumn), pp. 39–60.
Paris Declaration (2008) *Joint Declaration of the Paris Summit for the Mediterranean*, Paris, 13 July.
Philippart, E. (2003) The Euro-Mediterranean Partnership: unique features, first results and future challenges, *CEPS Working Paper*, 10, Centre for European Policy Studies, Brussels, April.
Scharpf, F. (2006) The joint-decision trap revisited, *Journal of Common Market Studies*, 44(4), pp. 845–864.
Schmid, D. (2002) Optimiser le processus de Barcelone, *Occasional Papers* 36, Institute for Security Studies of the European Union, Paris.
Schumacher, T. (2010) Transatlantic cooperation in the Middle East and North Africa and the growing role of the Gulf States, *Mediterranean Paper Series*, German Marshall Fund of the United States and Istituto Affari Internazionale, July.
Sénat (2009) *Situation de l'Union pour la Méditerranée. Communication de M. Robert del Picchia*, Reunión de la commission des affaires européennes, 31 March.
Soler i Lecha, E. (2008) *Proceso de Barcelona: Unión por el Mediterráneo. Génesis y evolución del proyecto de Unión por el Mediterráneo*, Fundación Alternativas Documentos de Trabajo, No. 28, Barcelona Centre for International Affairs. Available at http://www.falternativas.org/en/content/download/12018/369466/version/7/file/OPEX+28_definitivo.pdf (accessed February 2011).
Statutes of the Secretariat of the Union for the Mediterranean (2010) Available at http://www.europarl.europa.eu/meetdocs/2009_2014/documents/empa/dv/final_version_u/final_version_ufm.pdf (accessed 20 October 2010).

# France and the Union for the Mediterranean: Individualism versus Co-operation

## MIREIA DELGADO
Department of Politics, University of Liverpool, UK

ABSTRACT *Mediterranean policies are necessary for the European Union as a whole. However, a game of relations and interests exists, where different actors attempt to position themselves on the map. Some are able to occupy a central role, while others are pushed into the background. The Union for the Mediterranean (UfM) is a good example of this, with France being the visible head but not able to decide all questions of shape and content. This contribution addresses the kind of role played by France during the process of presentation and development of the UfM, focusing on the relative degrees of individualism and co-operation at different stages of this Mediterranean project.*

If there is a country concerned about its loss of influence, it is France. This situation is not new. The Gaullist obsession with maintaining France's *grandeur* still echoes and has roots in the heart of society and in the works of academia. No other country has generated such an extensive literature of this kind. The well-known book *La France est-elle encore une grande puissance?* (Boniface, 1998) is the first on a long list. Titles such as *La France qui tombe* (Baverez, 2003), *Adieu à la France qui s'en va* (Rouart, 2005), *L'Arrogance française* (Saint-Martin and Gubert, 2003) and *France in Crisis* (Smith, 2004) constitute a huge catalogue impossible to detail here. One may think that this is just a phase in academic history, a craze for this kind of literature. But this period has lasted for more than 20 years, having reached its height in 2005–06 under Chirac's unpopular government. At that point, as if it were an artistic movement or a school of philosophy, the movement was given a name: *déclinologie*.

*Déclinologues* helped Nicolas Sarkozy get elected to office in the hope that France would recover its former status. The current French president was determined to introduce changes in national and international politics, some of them announced during his electoral campaign. Meanwhile, Dominique de Villepin asserted that *déclinologues* were the 'prophets of doom' (Campbell, 2006). He underestimated the disenchantment of French society with its political class.

This is not a trivial point. The feeling that France is descending the pyramid of power is the reason for its foreign policy reorientation. Sarkozy's new government introduced bigger changes than usual: a new white book on security and defence (2008), a new white book on foreign policy (also 2008) and the *rapprochement* with the USA have been some of the measures taken to reinforce France's presence in the world in an effort to reverse this trend. A country that has succeeded in having a seat and a clear voice in all the important international organizations sees now that the 'newcomers' are, little by little, taking its place. It might not be France's fault. With its 632.834 km$^2$ and almost 64 million people, France is the second biggest country in the European Union, but it is very small compared to the new gigantic emerging powers. As a consequence, France has adopted the popular saying: you have to change with the times.

However, distribution of power and dynamics of influence are not the same in a Europe of 27 as in a Europe of 15 member states. Western Europe is no longer the centre. The East has been progressively gaining weight to the detriment of the West, and this trend is not likely to change in the foreseeable future considering that, at least for the moment, eastern Europe is the only possible direction for expansion. Sub-regionalization brings competition over EU funding and resources, affecting key areas of international relations such as commercial agreements, migration flows and security issues. Sub-regionalization also defines the nature of an external country's partnership with the EU; in other words, its prospects of obtaining 'advanced status' or of becoming a member state. This is part of the European Union's strategy towards its neighbours, which differs from region to region. The eastern one is much clearer, accession being the main incentive for eastern European countries. However, the EU does not have a clear model of regional integration for the Mediterranean (Escribano and Lorca, 2002: 2–3).

Consequently, this West–East dimension is a relevant factor to take into consideration together with the traditional North–South dimension (Barbé, 1998). There are two immediate consequences for France: first, decision-making is getting more complicated in a wider Europe which is expanding to the East; and, second, the traditional Franco-German axis has become unbalanced as Berlin has a stronger economy and a bigger territory under its direct influence. While Europe was (and still is) the natural area for French international projection, this Europe is now less willing to follow Paris without questioning why – a lesson that Sarkozy learned at the very beginning of his government when he put forward his 'Mediterranean Union' idea.

This lesson was particularly hard to learn given that the subject area was one that France considers to fall under its own umbrella: the Mediterranean region. Especially when even the Mediterranean partners failed to embrace the project that the French president proposed, and Germany insisted on a redesign of the whole initiative. It was a non-Mediterranean country that blocked France's ambition, and this was hard to accept.

Is France, despite the so-called decline, capable of taking a prominent role or even of becoming a regional leader? It was the *déclinologie* background that fostered the pro-activism toward the region, the Mediterranean being the card played by France to counteract its relative loss of influence. In a very small space of time, French diplomacy introduced a new framework for Euro-Mediterranean relations, convincing not only the Mediterranean partners but also the European member states to adopt it. Today this new initiative is largely institutionalized, and most of its features are based on French ideas, put forward during the presidential campaign. Furthermore, France managed to retain its co-Presidency of the Union for the Mediterranean notwithstanding the views of other EU actors that this role should have been transferred at the end of its EU Presidency. There is a strong French component in the UfM, but it was negotiation and not imposition that led to its establishment. Therefore, can France be regarded as a leader? The answer is not easy.

I will argue that a policy window existed,[1] although it was latent, and that France evolved from a form of leadership with strong components of unilateralism to acting as a genuine entrepreneur while keeping some features of the former style.

When Sarkozy proposed a Mediterranean Union, there was a window of opportunity that favoured the change. This policy window existed at all levels: international, European and national. At the international level, trends included the Mediterranean's progressive economic marginalisation (Khader, 2009: 181), a deteriorating political and social situation particularly in the great Middle East (Rapport Avicenne, 2007: 5–6) and a worsening of relations between Israel and Palestine, acting as an epicentre of regional frustrations (Rapport Avicenne, 2007: 10). At European level, the stagnation of the Euro-Mediterranean Partnership (EMP) and its disappointing results favoured the window of opportunity. Finally, at the national level, there were determining factors such as a feeling of losing influence in a more competitive and interconnected world, perceptions that France's Mediterranean policy needed strengthening (Khader, 2009: 181), the social unrest during Chirac's government and the tacit promise that the new president represented a change, a break with the past. Together, these factors created a propitious moment. All these events generated a type of policy window based on policy uncertainty[2] (Bicchi, 2007: 20), and this uncertainty gave impetus to what the *déclinologues* postulated.

However, although there were objective reasons that created this window of opportunity, it was not so evident at the time. In 2005 new goals were adopted within the EMP. The moment for further re-evaluation was to have been 2010, but France prompted the re-examination of objectives three years earlier than expected. Some countries are capable of bringing forward a policy window, France and the UFM being a good example. Electoral rhetoric and high-powered activity by French diplomacy expedited an event that was going to happen sooner or later, although not immediately. Other actors who had initially been unwilling to abandon the Barcelona Process for the UfM accepted after a process of negotiation. French action 'forced' the timing and adjusted it to France's convenience: the presidential change.

Regarding the French role during the process of early UfM development, I would define three different stages: a first one where the major feature was the exercise of traditional leadership based on a pronounced unilateralism, rigid proposals and a certain grandiloquent attitude embodied in the president, who also adopted an overly individualistic style. Then, a second stage where there were some signs of transition; although France's role changed in form, it kept the main features of a strategic leadership. Finally, it evolved toward a more co-operative role. Following the UfM's introduction, France became a genuine entrepreneur with some elements of strategic leadership.[3] Two factors, however, play a key role in understanding the background against which the UM/UfM initiative developed: a) France's government structure and the resources devoted to the Mediterranean and b) an Arab vs. Mediterranean dilemma.

The role of France in the design and subsequent evolution of the UFM is examined below, and in so doing, the contribution addresses three main questions. What kind of role has France played? Is its present co-operative attitude an adaptation to circumstances or a result of pressure from other actors? To what extent can France make its will prevail after having had to modify its original plan?

## A Powerful Machinery at the Service of the Mediterranean

The most important attribute of French foreign policy is the centrality of the *Elysée*. The president has the main role not only in foreign policy, but also in defence, a characteristic feature of the Fifth Republic (Vaïsse, 2009: 74). French foreign policy is directly linked to defence issues and military strategy. De Gaulle designated foreign policy to the presidential domain. The direct involvement comes from the interpretation of article 5 of the Constitution (Vaïsse, 2009: 17). Although under Sarkozy's government the Constitution was revised, in theory empowering parliament, in practice there have been no major changes in the domain of foreign policy, where the president retains the main role in what is likened by Vaïsse to a '*situation de monarchie absolue dans le domain de la politique étrangère*' [situation of absolute monarchy in the field of foreign policy] (Vaïsse, 2009: 74). In this sense, Sarkozy can be regarded as a traditional Gaullist.

The significant role of the president also has a negative counterpart: diplomats are given little leeway to act as French representatives in comparison with other countries. Furthermore, the Minister of Foreign Affairs is limited not only by the president but also because of the roles that other ministries have in the domain, although the co-ordination between them falls under the responsibility of the *Quai d'Orsay* (Vaïsse, 2009: 32–33).[4] Other institutions, such as the General Secretary and the Cabinet directors, have a variable role depending on the government.

France has the second largest diplomatic network in the world, after the USA. In 2008 France had 156 embassies, 17 permanent representations and 98 consulates.[5] It is interesting to note the importance that cultural diplomacy has for Paris, included as it is in all analyses and strategies. Cultural diplomacy is not just very active, but also highly institutionalized and the beneficiary of major resources. It is used as an important tool of soft power and influence. Therefore, 135 centres and cultural institutions have to be added to this outstanding amalgam of diplomatic centres, a number which has been constantly growing.

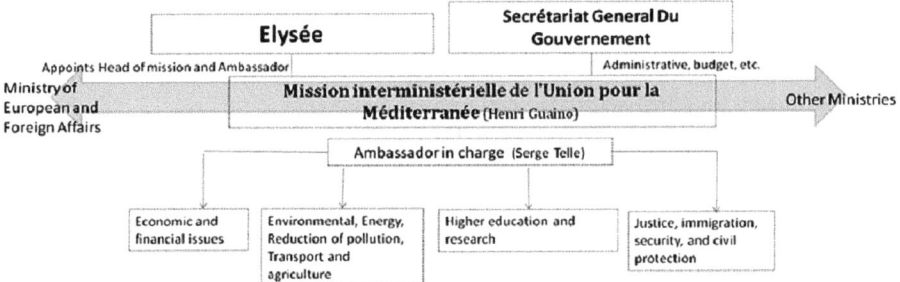

**Figure 1.** Organizational chart of the *Mission interministérielle de l'Union pour la Mediterranée*.
*Source:* Author's creation, with assistance from the French UfM team.

Turning to the specific case of the Mediterranean, it is included in one of the five geographical areas: North Africa and the Middle East. However, under Sarkozy's government, a new institution was set up: the *Mission interministérielle de l'Union pour la Méditerranée* (see Figure 1),[6] whose head is directly appointed by the president of the Republic (Art. 2) and which has its own personnel and resources (Art. 4). At the moment of writing, the team comprises nine members who are experts in different areas.[7] At the head of the *Mission interministérielle* is Henri Guaino, the ideologue of the Mediterranean Union and close advisor to the president. Serge Telle, former ambassador to the EMP, is the Ambassador in charge of the UfM. However, the Mission is not an institution of the *Elysée*. In terms of administrative issues, budget, etc., it reports to the *Secrétariat General du Gouvernement* (SGG). Nonetheless, the head of mission has an important political influence (Martín, 2009: 7), which highlights the presidential role, as the *Mission Interministérielle* is perfectly in line with the official tenets of French foreign policy.

The Interministerial Mission was created in order to achieve defined objectives: to ensure the success of the EU Presidency that France was holding in 2008, to launch the UfM projects agreed at the Paris summit and to co-ordinate the different French ministries that might be involved in different areas of policy. It is worth underlining that it is a unique institution. No other Mediterranean or non-Mediterranean partners have a similar body, a fact that the Mission regrets. In the opinion of a UfM officer,

> It would have been desirable that other countries install an interministerial task force similar to our own. Putting all the projects together in each of the countries, the task forces could have shared their initiatives with other countries. In the end, it is possible to ask whether that reflects the level of commitment in each country.[8]

Another body, created in December 2008, is the *Conseil Culturel de l'Union pour la Méditerranée*.[9] This is not linked to the Interministerial Mission. They are independent, complementary institutions that share the same facilities. The *Conseil Culturel* does not operate in the institutional framework of the UfM, but rather acts at the initiative of France as a network of people and projects. It reports to the prime minister and the *Sécretariat General du Gouvernement* and has been set up for just five years (Art. 1). Its objective is to promote the French cultural dimension in the Mediterranean together with the association *Marseille Provence 2013, capitale européenne de la culture* (Art. 2). In this sense, the *Conseil Culturel* has an important role in helping the 2013 event to succeed. It is composed of eight members from different French ministries and other qualified persons who are representative of civil society (Art. 3). At the moment, there are 22 such representatives from various areas who come from different countries, not only France. It also has a General Secretariat with four members (the general secretary and three project managers). The prime minister appoints the qualified persons at the suggestion of the president of the *Conseil Culturel* (Art. 3) who, in turn, is elected by the president of France (Art. 4). Renaud Muselier became its first president. This institution is again in line with the *Elysée* given that the president of the *Conseil Culturel* is appointed by the president of the Republic.

The operation of this agency is interesting from a strategic standpoint. The cultural dimension did not find a specific place among the UfM projects that were selected at the Paris summit. France wanted to include culture among the initial projects,[10] but cultural proposals failed to be prioritized. The *Conseil Culturel* may be seen perhaps as an agency created to fill the gap in an area vital for French interests. As mentioned above, cultural projection is one of its most powerful diplomatic instruments, promoting the culture of *Francophonie*. The Avicenne Report analyses how France might strengthen its presence in the world and bemoans the fact that today Arab television networks have moved from Paris to London, while highlighting the need to reverse this trend (Rapport Avicenne, 2007). The creation of the *Conseil Culturel de l'Union pour la Méditerranée* has created new channels for cultural projection, giving France a comparative advantage over other countries that have not followed its example and have not shown the same degree of activity. At the moment there are just six UfM project areas, but the cultural dimension could be added in the future.[11] The experience and network are already there, permitting the *Conseil Culturel* potentially to act as consultants if this dimension is ultimately institutionalized.

There is another important organization that has a certain influence on the Mediterranean policy of France: the Arab World Institute (AWI). By the nature of its activity, mainly in the area of cultural affairs and development co-operation, it is common for the director to be a politician or a diplomat who has worked in the Ministry of Foreign Affairs, although the Institute is not linked to any French ministry. The AWI serves to strengthen international relations and is also significant at the national level due to the large community of Arabs established in France; its public value has been recognized by the Interior Ministry and the Ministry of Foreign Affairs.

Meanwhile, France is active in all of the fora focusing on the Mediterranean, some of which were created at France's proposal. Just to mention some: the 5+5 Dialogue, the OSCE Dialogue for the Mediterranean, the two initiatives undertaken by NATO in the region (the Mediterranean Dialogue and the Istanbul co-operation initiative), etc. France is well aware of the vast amalgam of organizations already in place and has the 'willingness to put together and use a maximum of the initiatives which function to demonstrate the pertinence of the regional initiative (UfM) ... in a sense, the 5+5 is seen as a laboratory of experimentation. In other words, it allows us to test ideas, concepts and ultimately to know if they are applicable on a wider scale'.[12] Thus, overall, there is a gigantic and highly institutionalized machinery, either established by France or open to French influence.

## Arab vs. Mediterranean Policy: A Break with the Past?

When talking about France's foreign policy towards the Mediterranean, it is common to distinguish between an Arab policy and a Mediterranean policy. These are two sides of the same coin. Ever since the foundation of the Fifth Republic, France has maintained consistency toward the region. It was De Gaulle who founded the traditional Arab policy of France. Although during electoral campaigns presidential candidates tend to promise a break with the past, as in the case of Mitterrand (Vaïsse, 2009: 388) and Sarkozy, the truth is that once they arrive in the *Elysée*, it is continuity and not radical change that prevails. Economic interests and historical links are important factors that cannot be ignored.

France's Mediterranean policy is relatively recent. Paris has traditionally been a major player in the Mediterranean, but primarily through strong Arab relations. It was Mitterrand's presidency that developed French Mediterranean policy (Chérigui, 1997a).[13] Mediterranean policy is based on the progressive trend toward globalization and regionalization. However, there is a paradox as French understanding of 'Mediterraneanization' is also based on the concept of an 'individual Mediterranean' (Chérigui, 2000: 144). It includes states as well as non-state actors that share common values and whose nature is inclusive and multilateral. All of this is reinforced by the Mediterranean mythology that French diplomacy has used since 1930 (Henry, 2009: 21). French Arab policy, on the contrary, is exclusive as it is established on a monocultural basis, including only Arab countries. It aims for the reconstruction of a national discourse and has a strong component of bilateralism (Chérigui, 2000: 144).

According to Chérigui, since 1997 French Mediterranean policy has been the main diplomatic tool in its relations with the southern Mediterranean countries (Chérigui, 2000: 144). In 1996 Chirac made an attempt to re-establish the Arab policy, and for a short time both values coexisted although it did not last long (Chérigui, 1997b: 10–15). Nevertheless, this does not mean that the Arab policy has been abandoned. In fact, French policy towards the region is articulated around three axes: the Mediterranean, the Arab and the *Francophonie* (Chérigui, 2000: 145), which ultimately constitutes a powerful machinery to project influence toward the region.

My research suggests the existence of a hierarchy which includes Arab policy inside the Mediterranean policy where both dimensions coincide (see Figure 2). The Arab policy does not disappear, the two coexist, and what differs is the degree of emphasis as well as the institutional framework in which they operate. *Francophonie*, however, acts in a different dimension. It has a global nature since many of the actors are not specifically Arab or Mediterranean and therefore it contains some features that cannot exclusively apply to the region.

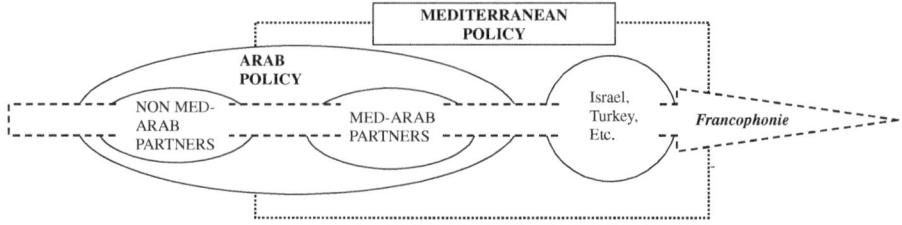

**Figure 2.** Mediterranean and Arab foreign policy in the foreign policy of France.
*Source:* Author's creation.

Usually, there is a simplification of the terms that leads to confusion when talking about these policies, identifying the Arab policy with a more pro-Arab approach in foreign policy, or linking the Mediterranean policy to a more pro-Israeli attitude. However, as shown by Chérigui, Mediterranean policy is not exclusively based on the degree of support for Israel.

It is against such a background that this contribution attempts to analyse the evolution of policy between the governments of Chirac and Sarkozy. Indeed, there are some elements of change in Sarkozy's approach to the Mediterranean that might indicate a reorientation and a break with the Gaullist tradition, although the magnitude of this change is not as remarkable as announced during his electoral campaign (Rubio, 2008: 4). Chirac was a Gaullist in the sense of defending strong Arab bilateral relations, independence from the United States and the existence of a national discourse. His regional policy was distinctly proactive, enhanced bilateral relations with Egypt being the cornerstone of this policy (Vaïsse, 2009: 404). In this sense, although France was fully involved in the Barcelona Process, Chirac invested great diplomatic resources in the Arab policy. The two policies coexisted, although it can be suggested that the accent was, at least for a while, on the Arab policy. This was not an arbitrary decision; on the contrary, the underlying strategy was to reinforce the role of France in the Mashreq while calling for a greater EU presence to counterpoise US influence (Chérigui, 1997b: 10). Chirac's strategy may have been doomed to fail, at least at the European level, given that just a few months before the creation of the EMP, it was already clear that a multilateral Mediterranean approach was the option chosen by the EU. So, even during this short period of time (1996) when both policies coexisted in France's relations with southern Mediterranean partners, in the end it was the 'Mediterraneanization' concept that became more prominent, although Chirac kept strengthening relations with the Arab partners.

When Sarkozy came to power, he entrusted the Mediterranean policy strategy to Henry Guaino, a self-confessed Gaullist. However, his discourse led one to think that he represented a big change. In the end, what prevailed was the traditional French consistency toward the region.[14] Some of his actions suggested a break with the past, like the return to the core of NATO or the pro-Israeli attitude which worried the Arab countries (Khader, 2009: 195). However, even this is debatable.

Although improvements in the relations between the Jewish state and Paris were a priority for Sarkozy, the president also called several times for Israel to leave the occupied Palestinian territories, as well as supporting the Palestinian right to statehood. It could be suggested that Sarkozy's aim was to become a reliable partner that could act as a mediator in an eventual peace process. He also engaged in determined activity to mediate between Israel and the Palestinians during the war in Gaza, condemning attacks and calling for the creation of a Palestinian state. Further evidence of continuity is that France acted to persuade Israel to accept the participation of the Arab League in all meetings of the UfM.

The Arab policy has not disappeared. Improved French relations with Israel do not necessarily imply a neglect of the traditionally strong ties with Arab partners. Indeed, French commitment to these partners was signalled by Sarkozy's announcement of a new permanent military base in Abu Dhabi which, for the first time since De Gaulle, represents a shift in military presence from Africa to the Middle East.[15]

However, under Sarkozy, greater emphasis is being given to the Mediterranean policy, as some features of French activity show: a regionalization trend, particularly in the original idea of the Mediterranean Union; the implementation of the ambitious UfM initiative for the Mediterranean region; the creation of *the Mission Interministérielle de l'Union pour la Méditerranée* and the *Conseil Culturel de l'UfM*; improved relations with Israel; and French presidential involvement in talks between the Palestinian Authority and Israel as co-president of the UfM and not as the president of France.

To conclude, despite some visible changes, there has been substantial continuity from Chirac to Sarkozy at the level of Mediterranean policy. Already under Chirac's government this policy was the main approach for French activity in the region, aimed at maintaining strong ties with Arab partners.

## The Road from Individualism to Co-operation

The Mediterranean Union was a project that has gone through different stages leading to its ultimate transformation. The transition from a 'Mediterranean Union' to the 'Union for the Mediterranean' was also accompanied by a transformation of the French discourse and role. After a first phase with strong traits of leadership, French diplomats adopted a collaborative attitude while trying to maintain their status as *primus inter pares*.

## An Unpopular Leadership

In 2007 Sarkozy announced his idea of a Mediterranean Union. As Barbé points out, it was not the first time that EU member states and the Commission had evaluated the state of Euro-Mediterranean relations. The first was in 2005 on the tenth anniversary of the Barcelona Process (Barbé, 2009: 21). However, it was in 2007 that a proposal questioned the essence of the EMP. What started out as a slogan in an electoral campaign was to transform the EMP. The Barcelona Process had given rise to a framework in which other partners could overshadow France (Rubio, 2008: 3). Consequently, Sarkozy was quick to announce a new initiative that involved a more active presence for Paris.

On 7 February, at a rally in Toulon, Sarkozy announced the need to prioritise the Mediterranean, concluding that the Barcelona Process had not met its objectives, which was in part because of the attention that the EU was giving to the East. Traditionally, Mediterranean policies have been viewed, in one respect, as a means of counter-balancing the eastern weight within the EU, and once more the Mediterranean Union seemed to serve this purpose by emphasizing the importance that the region has for Europe, or, in Sarkozy's words: '*car l'avenir de l'Europe est au sud*' [because the future of Europe is in the south].[16] However, he subsequently asserted the need to create a new entity similar to that of the European Union but excluding the EU from it. This Union would be inspired by Jean Monnet's approach, focusing on practical projects and leaving out political issues that might jeopardise their success. According to the president-to-be, the future of Europe was in the South but non-Mediterranean Europe had no place in it. This postulation was not only contradictory, but also quite surprising as Sarkozy himself presents France as '*européenne et méditerranéenne à la fois*' [European and Mediterranean at the same time]. This is one of the many contradictions that the proposition presented. How he was going to resolve them still is not clear, leaving the suspicion that there was no consistent design behind his electoral proposal. The project proposed in Toulon was very vague and idealistic, and for some analysts it recalled the 'civilizing mission' of France (Khader, 2009: 175), another element that maintains the individualism of French action.

It was not surprising that the project met with initial reluctance from the other European member states and the Commission, mainly because the initial proposal left out the EU. France's European partners reacted with equal disaffection, as did some of the Mediterranean partner countries.[17] The excessive unilateralism of the French president, together with the nature of the project that seemed to compete directly with existing EU Mediterranean policy, the exclusion of the non-Mediterranean European partners, the unilateralism in the design and presentation, and the grandiloquence with which it was outlined, all brought fears of a return to a more individualistic Paris.

During the following months, Sarkozy visited Algeria, Tunisia and Morocco with the intention of seeking partners to support his project, yet the proposal was still not clarified. The tone and content were ambiguous, raising many questions. The now-elected president undertook an intense level of activity which, together with the poor information offered in his public appearances, reinforced the impression that the French project was an attempt to re-nationalize the EMP (Schmid, 2009: 2).

During this phase of the project design, France presented certain characteristics of a unilateral and individualistic actor. This was a time of presentation, not interaction. The main actor, if not the only one, was France, and the rest were mere spectators. France was at this point a strategic leader. There were no components of entrepreneurship because there were no other actors in the design of the initiative.

The question remains whether the Mediterranean Union was really necessary and why it was designed to leave out the European Union. In Sarkozy's speech in Tangier, he acknowledged that progress had been made by the EU in relation to the Mediterranean. Furthermore, he admitted that the Barcelona Process and the European Neighbourhood Policy had contributed to a rapprochement between the two shores.[18] This tacit recognition suggests, once again, that the motivation was influenced by a national strategy. Leaving aside the EU, France presented itself as the mediator between Europe and the Mediterranean without having to compete with players such as Germany or the European Commission.

**First Signs of Transition**

The French attitude so far had generated strong rejection, forcing a change in style although not so much in content. The Ministry of Foreign Affairs through its secretary of state for European affairs, Jean Pierre Jouyet, tried to reconcile the Mediterranean Union with the EU (Schmid, 2009: 2). Differences became perceptible within the French team concerning the role that Europe might have in the new proposal. Ambassadors Huntzinger and Le Roy tried to reconcile tensions with France's European allies while promoting the initiative in the southern Mediterranean states. Countries involved in the project were asked to appoint a contact diplomat who would receive information on the Mediterranean Union and were asked to contribute with ideas. Up to 10 countries did so before the initiative was Europeanized (Soler, 2008: 12). Although France's discourse was less individualistic and more inclusive, 'French political leadership during these months was unquestionable, the motor of the initiative was in Paris alone, and the structures were not set in motion which allowed decisions to be taken jointly with its European and Mediterranean partners' (Soler, 2008: 12).[19] Interestingly, in this process of promoting the Mediterranean Union, the French ambassador in charge of the EMP[20] was not part of the team (Soler, 2008: 11), thus reinforcing the idea of separation between Europe and the Mediterranean that was indicative of the absent desire to include the EU. Moreover, during the whole process of negotiation, the intense activity of Sarkozy and of his close-knit team shows once more the centrality of the presidential role in French foreign policy.

In December 2010, when Sarkozy, Rodríguez Zapatero and Prodi met in Rome, some small compromises were made, without changing the core of the project. For Spain and Italy, the inclusion of the EU in the project was a compulsory condition for their support. The reluctance of Spain was even greater given that Madrid saw itself as the father of the Barcelona Process. The Italian and Spanish prime ministers ultimately supported the initiative on condition that the name was changed, it becoming the Union for the Mediterranean. The use of language is important because the earlier name, Mediterranean Union, implied a political union, while a Union for the Mediterranean limited the reach of the project (Gillespie, 2008: 277). According to the agreement reached with Spain and Italy, the EU was to be fully involved and the UfM was to complement and not replace the Barcelona Process.

At this point, France was in the driver's seat, not allowing other actors to become co-entrepreneurs (see Bicchi and Gillespie, this collection). The huge amount of diplomatic resources invested by the French and their weak disposition to co-operate with other EU actors continued to suggest that Paris was still acting as a strategic leader.

In the end, it was Germany that forced the whole readaptation of the UfM (see Schumacher, this collection). Germany even put on the table the threat of a veto. Although Spain and Italy had strongly supported the inclusion of the EU, it was not until the meeting in Hanover that its full participation was finally secured. For Sarkozy, this was a considerable step backward. The UfM was subordinated to the Barcelona Process, taking its name. From that moment on (until the Marseille meeting), it was 'Barcelona Process: Union for the Mediterranean'. However, there was a positive side to it as it now seemed clear that the project was going to go ahead.

The rapid development of the initiative was a result of intense diplomatic activity conducted mainly by France. This left some doubt as to exactly where the project was headed. French leadership, softened after the Rome meeting, might have been a strategy to force change onto other actors such as the European Commission and the northern countries, but this collaboration between France and Spain/Italy was calculated and in practice never existed very meaningfully; and when it did exist, it was still France that took the lead over future activity.[21] The moment when France turned into an entrepreneur was after the Hanover meeting. From that moment on, although France was still the visible head, it would be by means of co-operation with other actors that decisions would be taken.

**France as Broker Entrepreneur**

As of 13 March 2008, France started working in conjunction with the European Union and other actors. However, the Presidency of the EU would test France's readiness to co-operate with others. Typically, presidencies are used by member states to promote their own priorities; consequently, it represented a new opportunity for French *protagonismo*. Sarkozy included the proposed UfM in a privileged position on the agenda.

This period was characterized by some unexpected events that modified the original plan: the conflict in Georgia, the global financial crisis and the Irish rejection of the Lisbon Treaty. The excessive personalization of French foreign policy became apparent during the whole Presidency; although this does not necessarily mean that it was counterproductive, for the Presidency had to face some challenges that required full involvement. Nevertheless, as Lequesne and Rozemberg assert, 'A certain lack of co-operation with other member states, particularly Germany, has been the price to be paid for such activism' (2008: 6–7). They continue, 'The energy of the young French president was also associated with what Brussels has traditionally considered to be France's main fault – arrogance' (2008: 8). However, as both authors finally conclude, the scope of the events that followed and the strong implication of French diplomacy helped to soften the 'messianic self-perception' (2008: 9).

The initial EU Presidency programme included four major topics that reflected French national priorities, plus one additional topic: immigration, defence, climate change and energy, and finally agriculture.[22] This agenda was modified due to the events mentioned above, and also because of the inclusion of the UfM. Moreover, Morocco's advanced status was included in the list of priorities.

The ambivalence between clear allusions to leadership and a more co-operative role was constant before and during the French EU Presidency. Again, Lequesne and Rozenberg point to this feature of Sarkozy's discourse in the claims made that during his presidency he would emphasize the common immigration policy, defence, energy and the environment. This would not have been so striking in the presentation of a proposal had the French statement not been made on 8 January 2008 while Slovenia, preceding France in the EU Presidency, was presenting its own presidential plans. The caustic and ironic Slovenian response was that its own presidency would not be 'so great' but more centred on specifics. Jouyet had to clarify the situation and quickly tried to soften the unilateralist image of the French president (Lequesne and Rozenberg, 2008: 17).

France's Presidency began with an important change: Alain Le Roy, the ambassador in charge of the UfM, was appointed as the new deputy secretary-general for peacekeeping operations at the United Nations. Le Roy knew about his UN assignment several months beforehand; however, media suggested that differences between Guaino and Le Roy were behind this decision (EurActiv, 2008).[23] The natural choice to replace Le Roy was Serge Telle, but it was Guaino who was appointed as head of the mission and Telle kept his existing post. The appointment of Guaino, a close advisor to the president, ensured that the new Interministerial Mission would stay in line with the presidential strategy which, yet again, shows the centrality of the *Elysée* in terms of foreign policy.

In order to guarantee the success of the Paris summit on 13 July, French diplomacy had to deploy all its resources. During the preparation of this meeting, France acted as an entrepreneur, contacting other states to ensure their participation and contribution to the UfM. The presence of the Syrian president at the summit and the rumours of an eventual resumption of Syrian–Israeli peace negotiations provide a good example of the kind of entrepreneurship action carried out by Paris in order to ensure success. Almost all the countries attended: only the Libyan leader and the kings of Morocco and Jordan refused to attend. The absence of King Mohamed VI was surprising given the good relations between France and Morocco. The summit ended with a sense of continuity with the Barcelona Process but with a greater role for the intergovernmental dimension, to the detriment of the European approach. Such a design was compatible with the original idea of the project. In this game of relations, French diplomatic effort and capacity for action in the region exceeded by far the mobilization of resources by other countries included in the project.

It was in Marseille that the Barcelona Process: Union for the Mediterranean left behind the reference to Barcelona. Holding this foreign ministers' meeting was not as easy as holding the one in Paris.[24] To begin with, Israel had opposed the idea of Arab League participation in all UfM meetings as an observer. Also, during the months that preceded the ministerial meeting, several countries competed to host the UfM Secretariat. It was a race of diplomatic contacts and declarations. France itself presented the candidature of Marseille for the 'capital' of the Mediterranean, but it would have been surprising had the Secretariat been allocated to that city, considering that France was already in the co-Presidency. In principle, the Secretariat should have been placed in a southern country in the interests of co-ownership;

however, southern countries could not agree on a candidate and it was Barcelona that was finally chosen. Spain gained the physical placement and, in return, allowed France to remove any reference to the UfM that sounded too Spanish. Nevertheless, following Soler's line of argument (Soler, 2009: 172), 'to ensure that the Ministerial (meeting) ended with agreements, and with the aim of satisfying the maximum number of countries, the declaration of Marseille is an exercise in balance and ambiguities that significantly affects the Secretariat'.[25] As a result, two things stand out from the Marseille meeting: on the one hand, France consolidated its role as a genuine entrepreneur, making concessions and showing a less rigid and more co-operative attitude. On the other hand, France did not renounce the role of *primus inter pares*. Paris got back the name and the institutional design consolidating the intergovernmental approach.

This intergovernmental dynamic already seemed to be in France's initial proposal of the Mediterranean Union. By giving all actors the same influence, the initiative involved a strong intergovernmental approach where states were not subordinated to any supranational body. It was the EU that sponsored the EMP, whereas, as Aliboni says, 'the primary 'driving-force' of the UfM is to overturn this unequal balance of power with a shift towards an organizational structure based on the principle of "parity"' (Aliboni, 2009: 3). This feature was present from the first day the UM was announced. All states would be operating under equal conditions and would have the same weight. In addition, in the case of the European states, they would not be subjected to the European Union; all countries would be given more leeway, gaining independence. With the inclusion of the European Union, the intergovernmental approach was not so clear-cut, but after the Marseille meeting the new UfM structure consolidated this dynamic, weakening the role of the EU compared to the function that it had in the original EMP.[26] All in all, a very similar model to that proposed at the beginning.

The French Presidency of the EU, in general terms, was a success despite the setbacks (Lequesne and Rozenberg, 2008). However, as noted above, it suffered from a certain paternalism marked by the great activism and rhetoric of the president. However, it is through negotiation and compromises that France has acted in the UfM, and not through imposition. In this regard, French diplomacy became an entrepreneur working hand-in-hand with other countries and with the European Commission.

The invasion of Gaza caused the first great crisis of the UfM. For almost a year, ministerial meetings were postponed indefinitely. French diplomacy, with the president at the head, used all its influence, sometimes acting in parallel with the EU Presidency. This hyperactivity was regarded, again, with suspicion, especially when Paris began to insist on maintaining its co-Presidency of the UfM beyond the six-month term of its EU Presidency.

Overcoming the obstacle created by Israel's move against Gaza proved almost impossible. Paris mobilized to condemn the invasion but, at the same time, insisted on the security of Israel. This balance was coherent with French foreign policy and also with France's role as a co-president of the UfM. Some informal contacts took place, but the Arab embargo made it impossible to continue normal activity. In late

November 2008 there were some signs of progress: it was announced that the Jordanian Ahmad Masadeh would head the Secretariat and ministerial meetings were resumed, but not everything was resolved. It became clear that side-stepping the Arab–Israeli conflict was simply not possible.

The Spanish Presidency of the EU early in 2010 was the moment to test French entrepreneurship.[27] Spain was the first Mediterranean country, after France, to hold the Presidency of the European Council. Furthermore, the entry into force of the Lisbon treaty brought a new complexity to the structure of Euro-Mediterranean relations that required even more so a spirit of entrepreneurial action. The main Spanish task regarding the Mediterranean was the UfM summit scheduled to take place in June 2008. It did not happen; however, France and Spain worked together to try to make it happen. The unsatisfactory outcome was not because of a lack of understanding between the two neighbours but because of the complicated circumstances that once again affected the Mediterranean initiative. Co-ordination between France and Spain was stressed not just for the Presidency of the EU but also for the development of projects as Madrid was seen by Paris as an active partner with great interests in the region, serving the UfM.[28] Some elements of competition may appear, but at the moment they have not seriously affected co-operation between the two countries. Consequently, there is some reason to suggest the progressive establishment of co-entrepreneurship. It seemed implicit in French support for Spain to succeed France in the co-Presidency of the UfM, which some observers were expecting to happen (Aliboni, 2009: 6).[29]

In this third stage, France combined both a high and a low profile in its entrepreneurial action. Its high level of activity during the Gaza conflict showed commitment to the region but it also overlapped with the role of the European Union and the EU president, creating some confusion. In this sense, it can be argued that there was a sort of regression. While it may have been justified by its function as co-president of the UfM, co-ordination between the EU Presidency and France on certain subjects did not seem to be very effective, showing some trends of the initial individualism.

This attitude contrasts with a less intrusive profile within the UfM, leaving room for other actors to take the lead in certain areas and project developments. France now seeks co-ordination with all the partners and even asks for greater commitment on the part of certain southern Mediterranean partners who, so far, have not been especially proactive in this regard. Within the UfM, it seems that France acts as a genuine entrepreneur, creating a political space for co-entrepreneurship. However, resources and involvement of other partners will also be a necessary condition for the successful development of this co-entrepreneurship.

## Conclusions

Is France a leader of the Mediterranean initiative? The answer is 'relatively'. There is little doubt that now more than ever the Mediterranean is associated with Paris. The active role that France has conducted brought a kind of reaction that has not always favoured French interests, but surely has put France back in the headlines.

The first phase suggested leadership in the full sense of the word. It was also effective considering that France proved capable of dragging into the initiative the two biggest Mediterranean partners (Spain and Italy) and, in the end, forcing a reorientation of the Mediterranean policy of the EU. It involved a game of approaches to other actors but with only small concessions that might be seen as a prerequisite for dealing with Germany. This was the moment when France was the unique actor and the unquestionable leader.

The later stage, marked by an increase in co-operation, could lead one to conclude that France has softened its role, and to a certain extent this is true, at least in relation to traditional notions of unilateralism and rigidity in policy making. However, if we look at what, how and when, we can see the extent of French influence. Clearly, France knew how to use its Presidency of the European Council. The timing was propitious, and it fell between the EU presidencies of two relatively small countries which strengthened its capability for action. This might explain why France, at this point, acted more as the guardian of the initiative and began to adopt a more co-operative attitude. Paris discovered that traditional leadership could not be effectively exerted without including Germany or the northern countries. Nor could it count on Spanish support if there were no space for collaboration. Thus France became an entrepreneur. Subsequently, this second stage was based on influence and co-operation, although with some elements that might indicate a certain degree of involution in their co-operative action and a desire to keep a certain degree of influence in the formula of *primus inter pares*.

The question remains whether the leadership exercised by Paris was earned entirely by merit or was a product of the vacuum left by other member states. Some of the factors that have influenced the preponderance of Paris are directly attributable to France, which invested greatly in diplomatic and technical resources. However, the lack of strategy and resources from other countries, particularly from the moment that the Paris summit took place, could be a further determining factor in French predominance. It shows that not all states are willing to assume a proactive role. It is noteworthy that only France has an interministerial team, an institution that one might have been expected to see emulated at least by other countries that aspire to play a more significant role in the region.

To conclude, Sarkozy's commitment is to continue strengthening the Mediterranean policy, even more than his predecessor; and his initial idea was a Mediterranean region based on the parity of its members. The concept of parity is hardly compatible with strategic leadership. These two elements are indicative that despite the strident beginning of the initiative and some elements in the style of the president, the entrepreneurial spirit was already part of the French plans for the *Mare Nostrum*.

## Acknowledgements

I would like to thank Richard Gillespie and Federica Bicchi for their comments, suggestions, notes and, most important, their support. I also very much welcome all the ideas offered by Hakim Darbouche, Iván Martín and the anonymous reviewer. The comments of the participants in the workshop 'The UfM:

THE UNION FOR THE MEDITERRANEAN

Continuity or Change in Euro-Mediterranean Relations?' at the LSE on 10 May 2010, and those made at the WOCMES Conference in Barcelona on 19–24 July 2010 were very valuable. I want to thank Colt Segrest for his help on this trilingual project. Finally, a very special mention of all the team of the *Mission interministerielle de l'Union pour la Méditerranée* and the *Conseil Culturel de l'Union pour la Méditerranée*, from the secrétariat to all the officials and diplomats.

I would also like to thank the French UfM team for contributing to the design of Figure 1. If there is any inaccuracy, it is mine.

## Notes

[1] A window of opportunity, as Bicchi explains, is a 'confluence of environmental conditions favoring the rise of a particular issue on the political agenda at any given moment' (Bicchi, 2007: 19).

[2] Bicchi argues that 'cognitive uncertainty occurs when patterns of domestic politics are shaken, for instance, by the perception of a new challenge' and says: 'challenges originating from outside national borders can thus have a disruptive impact on national interests, as they challenge engrained policy preferences.' This kind of policy window was at the core of the EMP (Bicchi, 2007: 20).

[3] For further details of the typology, see Bicchi (this collection).

[4] As an example, in 2008 just 47 per cent of the total budget and half of the human resources for the international action of France were for the Ministry of Foreign Affairs. This presented two problems: lack of coherence in some actions and the necessity of taking into consideration the interministerial dimension of the international projection of France (Juppé and Schweitzer, 2008: 104).

[5] Data published on the website of the Ministry of Foreign Affairs. Available at http://www.diplomatie.gouv.fr/fr/ (accessed 15 October 2010).

[6] Decree no. 2008-1188 of 14 November 2008 on the creation of the *Mission interministérielle de l'Union pour la Méditerranée*.

[7] Interview with a French UfM official, October 2010.

[8] Interview with a French UfM official, October 2010. My translation.

[9] Décret no 2008-1277 of 8 December 2008 on the creation of the *Conseil Culturel de l'Union pour la Méditerranée*.

[10] Interview with a French official of the Conseil Culturel de l'UpM, October 2010.

[11] At the time of writing, there is no sign of a cultural project being included in the framework of the UfM.

[12] Interview with a French UfM oficial, October 2010. My translation.

[13] Chérigui analyses the evolution from the Arab to the Mediterranean policy from the perspective of regional leadership.

[14] In all the interviews that I have conducted with academics, diplomats and officials, when addressing this topic, there has been a consensus: despite some differences of approach, there is a basic continuity in French Arab-Mediterranean relations.

[15] Many thanks to an official of the Conseil *Culturel de l'Union pour la Méditerranée* for this observation in relation to this debate of Mediterranean vs. Arab policy (22 October 2010). For further information (see Tran, 2009).

[16] Speech by Nicolas Sarkozy, Toulon, 7 February 2007.

[17] See Del Sarto and Schlumberger (this collection) for analysis of the reactions of the southern countries.

[18] Speech by Nicolas Sarkozy, Tangier, 23 October 2007.

[19] My translation.

[20] François Gouyette, who would be replaced by Serge Telle.

[21] On the Spanish role and collaboration with France, see Gillespie (this collection).

[22] Work programme, available at: http://www.eu2008.fr/webdav/site/PFUE/shared/ProgrammePFUE/Programme_ES.pdf

[23] Union pour la Méditerranée: Guaino reprend la main. EurActiv.fr. 10 September 2008.

[24] On Israel, see Del Sarto (this collection).

[25] On the Secretariat, see Johansson-Nogués (this collection).

[26] On regionalis/regionalization and the EU, see Holden (this collection).

[27] On Spain, see Gillespie (this collection).

[28] A high level of co-operation and co-ordination between the two countries was confirmed to me in interviews with a Spanish official (July 2010) and French UfM officials (October 2010).

[29] At the time of writing, it is still not clear how the northern co-Presidency will operate following the launch of the European External Action Service; consequently a Spanish co-Presidency cannot be taken for granted.

## References

Aliboni (2009) The Union for the Mediterranean. Evolution and prospects. Document *IAI 09, 39e* (Istituto Affari Internazionali).
Baverez, N. (2003) *La France qui tombe* (Paris: Perrin).
Barbé, E. (1998) Balancing Europe's eastern and southern dimensions, in: J. Zielonka (Ed.) *Paradoxes of European Foreign Policy* (Leiden: Kluwer).
Barbé, E. (2009) La Unión por el Mediterráneo: De la Europeización de la política exterior a la descomunitarización de la política mediterránea. *Revista de Derecho Comunitario Europeo*, 32, pp. 9–46.
Bicchi, F. (2007) *European Foreign Policy Making toward the Mediterranean* (Basingstoke: Palgrave Macmillan).
Boniface, P. (1998) *La France est-elle encore une grande puissance?* (Paris: Presses de SciencesPo).
Campbell, M. (2006) French swing behind Sarko's revolution, *The Sunday Times*, 12 February.
Chérigui, H. (1997a) *La politique méditerranéenne de la France: entre diplomatie collective et leadership* (Paris: L'Harmattan).
Chérigui, H. (1997b) Maghreb et Machrek dans la politique étrangère de la France depuis l'apres guerre du Golfe. *Revue Banquet*, 11(July–December), pp. 51–74.
Chérigui, H. (2000) La politique méditerranéenne de la France: un instrument de leadership dans l'espace regional, in: J. R. Henry & G. Groc (Eds) *Politiques méditerranéennes entre logiques étatiques et espace civil* (Karthala: Annuaire de l'Afrique du Nord).
Escribano, G. & Lorca, A. (2002) El Mediterráneo: Frontera Sur de la Unión Europea, in: E. Palazuelos & M. J. Vara (Eds) *Grandes Áreas de la Economía Mundial* (Barcelona: Ariel).
*EurActiv.fr* (2008) Union pour la Méditerranée: Guaino reprend la main, *EurActiv*, 10 September. Available at: http://www.euractiv.fr/presidence-francaise-ue/article/union-pour-mediterranee-guaino-reprend-main-001048
Gillespie, R. (2008) A 'Union for the Mediterranean' ... or for the EU?, *Mediterranean Politics*, 13(2), pp. 277–86.
Henry, J. R. (2009) Mediterranean policies of France, in: J. R. Henry & I. Schäfer (Eds) *Mediterranean Policies from Above to Below* (Baden Baden: Nomos).
Juppé, A. & Schweitzer, L. (2008) *La France et l'Europe dans le monde. Livre blanc sur la politique étrangère et européenne de la France 2008–2020* (Paris: La documentation française).
Khader, B. (2009) *L'Europe pour la Méditerranée. De Barcelone à Barcelone (1995–2008)* (Paris: L'Harmattan).
Lequesne, C. & Rozenberg, O. (2008) The French Presidency of 2008: the unexpected agenda. *Swedish Institute for European Studies*. SIEPS 2008:3op.
Martín, I. (2009) Las prioridades de la presidencia española de la UE en el Mediterráneo: ser y deber ser. *ARI No. 166/2009* (Madrid: Real Instituto Elcano).
Rapport Aviccene (2007) Maghreb – Moyen-Orient. Contribution pour une politique volontariste de la France. 23 April. Available at: http://www.ifri.org/files/Moyen_Orient/Avicenne_DBauchard0407.pdf
Rouart, J. M. (2005) *Adieu à la France qui s'en va* (LGF. Le livre de poche) (Paris: Grasset).
Rubio Plo, A. R. (2008) La política mediterránea de Francia: del imperio latino de Alexandre Kojève al neogaullismo de Henri Guaino, *ARI Paper, 86* (Madrid: Real Instituto Elcano).
Saint-Martin, E. & Gubert, R. (2003) *L'Arrogance Française* (Paris: Balland).
Schmid, D. (2009) Du processus de Barcelone à l'Union pour la Méditerranée: changement de nom ou de fond? Dossier La Méditerranée. *Questions Internationales* (March–April).

Smith, T. B. (2004) *France in Crisis. Welfare, Inequality and Globalization since 1980* (Cambridge: Cambridge University Press).
Soler i Lecha, E. (2008) Proceso de Barcelona: Unión por el Mediterráneo. Génesis y evolución del proyecto de Unión por el Mediterráneo, *Documento de Trabajo 28/2008*, Fundación Alternativas y Fundación Cidob. Observatorio de política exterior española.
Soler i Lecha, E. (2009) La presidencia Francesa de la UE y la Unión por el Mediterráneo. ¿Una Europeización forzada?
Tran, P. (2009) France cultivates stronger ties in Gulf, *Defense News Journal*, 16 November. Available at http://www.defensenews.com/story.php?i=4376846
Vaïsse, M. (2009) *La puissance ou l'influence? La France dans le monde depuis 1958* (Paris: Fayard).

# Adapting to French 'Leadership'? Spain's Role in the Union for the Mediterranean

RICHARD GILLESPIE
Department of Politics, University of Liverpool, UK

ABSTRACT  *One might have expected Spain to have been less than enthusiastic about President Sarkozy's Mediterranean initiative of 2007–08, given that it implied criticism of the Barcelona Process in which Spain had been so prominent. Yet Franco–Spanish rivalry has not come to the fore as a result of it. Acceptance, adaptation and eventual support have been the keynotes to the Spanish response, notwithstanding some ongoing differences in outlook between the two countries. A political accommodation has been reached, with France guaranteeing Spain a visible, though not substantial, role in the Union for the Mediterranean (UfM). This builds upon earlier Franco–Spanish collaboration in the context of EU Mediterranean Policy, as well as close bilateral co-operation in various other policy domains, including migration and counter-terrorism. Collaboration with France also finds explanation in the wider EU context where it is now more difficult for any single country (even France) to achieve strategic policy changes. Besides analysing Spanish reactions to the assertion of French 'leadership' in the Mediterranean, this contribution considers Spain's potential to exert influence on the UfM in the future by looking at its EU Presidency during the first half of 2010.*

While France occupied centre-stage at the launch of the Union for the Mediterranean (UfM) in July 2008, Spain quietly repositioned itself in the wings. It had begun to adapt to a new role that generally would be quite supportive of France, but also aimed to maintain Spanish influence and ensure that what was originally a unilateral French initiative would not cause lasting damage to the EU Common Foreign and Security Policy. This process took place during a first phase in the emergence of the new policy framework, running from the promotion of the 'Mediterranean Union' (UM) idea in 2007 to its conversion into the Union for the Mediterranean in mid-2008. A second phase of adaptation ensued, with Spain returning to the stage to join France and regain visibility within the UfM. Today, Spain remains a significant actor in Euro-Mediterranean relations and, as an advocate of 'more Europe', has helped curb the tendency towards pure intergovernmentalism in President Sarkozy's Mediterranean policy (Barbé, 2009). The arrival of a Spanish Presidency of the EU during the formation period of the UfM was welcomed by many countries as an opportunity to consolidate the UfM on a basis that enhanced rather than broke the Euro-Mediterranean Partnership (EMP). Yet no change of orientation took place during those six months, leaving doubts about Spain's capacity for taking autonomous initiatives.

Despite historical rivalry between the two countries (particularly in North Africa), active competition to shape the EU Mediterranean agenda has not dominated Franco-Spanish relations in the Euro-Mediterranean context. While latent rivalry still surfaces on occasion – from France's blocking of EU support for Spain during the latter's spat with Morocco over Parsley Island in 2002 (Cembrero, 2006) to France's determination to see the 'Barcelona Process' prefix removed from the name of the UfM in 2008 – Spain's main concern has been to work *with* France (and other actors) to ensure that certain shared policy objectives are pursued more vigorously in the Mediterranean. Spanish enthusiasm for bilateral collaboration with France was discernible within the EMP well before the French initiative of 2007 and it was not killed off by the unilateralism that characterized President Sarkozy's *démarche*.

However, the interest in working together has been complicated by differences in role perception in relation to the Euro-Mediterranean space. If during the 1990s Spain acted as an 'entrepreneur', looking to build an effective coalition within the EU in order to enhance the latter's Mediterranean Policy through establishing a new multilateral consensus (Bicchi, 2007), France under Sarkozy has attempted to exercise 'leadership' in a more unilateral, less compromising fashion, reliant upon a major investment of *national* effort and resources (see Delgado, this collection), while using intergovernmental side-deals to gain support over specific issues, both within and beyond the EU.[1] Since 2007, France has been obliged by resistance from other actors to compromise with European partners, but much of its design has survived and the UfM agenda continues to bear a strong French imprint. Indeed, senior presidential advisors in the Elysée still insist that the UfM is purely intergovernmental and some EU countries remain concerned about French or possibly French–Spanish dominance.[2] The French approach has been modified but not abandoned,[3] thus leaving uncertainty as to whether Spanish co-leadership might be entertained by Sarkozy.[4] While seeking to function as co-leader, Spain is well aware that its economic and diplomatic resources pale by comparison with France's, potentially making it look more like a 'junior partner'. This notion is understood here to imply a defined influential role for Spain within the context of ongoing overall strategic leadership by France, whereas the concept of 'co-leadership' is seen as denoting a more fluid relationship between France and Spain: not an essentially equal one, but one in which there is no fixed hierarchy.

There are certainly reasons to doubt whether Spain will be able to play a co-leadership role. Besides the obvious economic constraints, made worse by the country's particularly deep recession, Spain is inhibited by a current absence of strategic vision, political ambition and belief in Mediterranean region-building, compared with its diplomacy in the latter years of the twentieth century. Although former foreign minister Miguel Ángel Moratinos referred to Spain's ongoing 'vocation to lead in the Mediterranean' (Moratinos, 2009: 16), Spanish objectives within the UfM have largely been restricted to the institutional domain – providing the headquarters for the permanent Secretariat in Barcelona – rather than seeking to shape the policy agenda or advancing a coherent strategy to pursue long-range objectives. While securing the Secretariat kept Spain on the Euro-Mediterranean map, it has not provided a platform for Spanish influence, given that no Spaniards occupy its key posts.[5]

To capture the dynamics of recent Hispano-French relations in the Euro-Mediterranean context, this contribution will begin by demonstrating that co-operation between these neighbours was at least as prominent a feature as any latent rivalry during the period of the Barcelona Process in 1995–2008. Proceeding to examine Spanish responses to the French bid for Mediterranean 'leadership' from 2007, Spain's reasons for embracing the UfM will be analysed. Thereafter the degree of congruence between French and Spanish positions will be examined. Finally, in the aftermath of its EU Presidency in 2010, Spain's current and future status within the UfM will be considered. Is there sufficient chemistry between the two neighbours for a stable partnership to develop? If so, is Spain's role now that of a 'junior partner', or could the balance shift further, allowing for major Spanish initiatives to prosper?

## Changing Places? Spain and France in the EMP

The time-frame for assessing national influence in the shaping of Euro-Mediterranean structures of co-operation is crucially important. Most studies of the UfM go back to the creation of the EMP in 1995. While this seems reasonable, given that 1994–95 and 2007–08 were periods of institutional innovation and policy development in Euro-Mediterranean relations and the UfM is an adaptation rather than a replacement of the EMP, too much emphasis on this period and on the high-profile events that launched each enterprise could give the impression that Spain has been replaced by France as the main driver of Euro-Mediterranean politics. However, a more historical perspective extending back to the 1970s reminds us of France's original pre-eminence in the context of a much smaller EC, prior to the Iberian accession, which also reveals the brevity and infrequency of those windows of opportunity when national impetus and influence really contribute to policy renewal. Spain worked very effectively to achieve influence over EU Mediterranean policy in the early 1990s (Gillespie, 2000: 134–77), but its rise to pre-eminence was facilitated by a diversion of French interest towards central-eastern and south-eastern Europe.

Spanish influence is clearly identifiable in this period, with national ideas from the late 1980s and early 1990s finding reflection in the eventual design of the EMP; but this was achieved through Spain developing the skills of an entrepreneur rather than becoming a leader, and it involved a substantial amount of compromise by Spain over the new framework for Euro-Mediterranean relations throughout the formative process. Thereafter, Spain acquired lasting prestige by hosting the Barcelona

Conference in 1995; it continued to prioritise the Mediterranean (more than any of the larger EU member states) over the years that followed and brought new policy areas onto the Euro-Med agenda during the Spanish EU Presidency of 2002. Over the same period, however, the Euro-Mediterranean region-building component that had featured in Spanish strategic thinking gradually lost momentum against a background of failure in the Middle East peace process,[6] while Europeanizing dynamics were reinforced (Bicchi, 2006), primarily through the parallel introduction of the European Neighbourhood Policy (ENP).

Despite initial concerns about the eastern neighbourhood being prioritized at the expense of the South, Spain soon found in the ENP an opportunity to achieve increased interdependence between (at least some) North African countries and Europe through its bilateral action plans, while also trying (with more disappointing results) to take advantage of its position as an EU frontier state to benefit from its cross-border co-operation (CBC) programme. Ultimately, with the more ambitious region-building plans frustrated by the persistence of conflict in Europe's southern neighbourhood and the endurance of structural impediments to co-operation, the multilateral institutions of the EMP which Spain had lobbied for in the early 1990s began to decay, though relatively low-key regional Euro-Med programmes survived. Faced with disagreement within the EMP, Spain promoted new forms of sub-regional collaboration, such as the policing of northward migration across the western Mediterranean, with some success. However, it was not able to revitalize the Partnership as a whole, given that the most ambitious goals of the EMP required constructive interactions between the different 'baskets' of the Barcelona agenda and the resolution of regional conflicts.

The limits to Spanish influence through the EU become more understandable if one also considers Spain's own regional policy, which earlier had been seen as an essential aspect of a strategy involving triangulation between Spain, the EU and the Mediterranean. Aspiring to develop a comprehensive ('global') regional strategy since the 1980s, though with priority given to the Maghreb, Spain was disappointed in its endeavour to improve relations simultaneously with all the Mediterranean partner countries. Efforts to achieve balanced relations faltered in the face of strong regional rivalries and persisting conflicts, not least those between Morocco and Algeria in the Maghreb; and the attempt to extend Spain's own presence initially proved more difficult in Algeria than Morocco for a variety of historical, political, economic and cultural reasons (Gillespie, 1995: 159–77). Once the ENP had been adopted, modifying the framework of Euro-Mediterranean relations, it became harder still for Spain to retain a region-building perspective.[7] Indeed, as its initial concerns about the ENP faded, Spain quickly 'internalized the philosophy behind the policy and tried to make this as sensitive as possible to the interests of Spain' (Soler, 2008: 4).

Within ten years of the launch of the Barcelona Process, the meagre results of Spain's overarching regional policy were giving rise to a new emphasis on reinforcing bilateral relations with those individual countries (notably Morocco and Turkey) that expressed enthusiasm for closer co-operation; yet relations with Algeria were hampered by disputes over energy issues and even relations with

Morocco were troubled at times. Spain's pattern of Mediterranean relations began to resemble more closely those of the wider EU, as the country became more Europeanized and as the varying responses of southern partners to the ENP themselves added to unevenness in Euro-Mediterranean relations. Indirectly, such developments affected Spain's ability to exert influence through the EMP, given the unanimity required for all major Partnership decisions. Its diminished standing was seen at the summit in Barcelona in 2005 to mark the tenth anniversary of the Partnership. Co-organizing the event with the then British Presidency of the EU, Spain made great efforts to mobilize participation at the highest level, only to lose face when most of the southern leaders stayed away from the event, at which there was insufficient agreement for a common final declaration to be adopted.

In this context of an ailing EMP which it wanted to revitalize and reinforce, Spain showed particular interest in collaboration with France. The tendency was facilitated by Spain's return to the European mainstream, post-Iraq, following the election of José Luis Rodríguez Zapatero's Socialist Party in the general election of March 2004. Although Spain under Aznar had helped give the EMP fresh political impetus through the Valencia Conference of 2002 (Gillespie, 2002), its image as a champion of the values of Barcelona had been damaged in the Arab world by its role in the Iraq War, leaving Spanish regional influence in question (Núñez Villaverde, 2005: 105–6). The return of a Socialist government was accompanied by pledges to relaunch Spain's own Euro-Mediterranean policy, partly through holding the unprecedented summit in Barcelona (Soler, 2008: 2), yet without any big new idea being proposed at policy level to revive enthusiasm in the South. While organizing this event with the UK, the Spanish government looked mainly to France as a potential ally, essential for the effective pursuit of greater ambition in EU Mediterranean policy. A French–Spanish–Italian 'non-paper' produced ahead of the summit identified a number of areas where joining forces was counselled by a shared desire to be more influential in the Mediterranean, particularly vis-à-vis the Maghreb.

In this period, foreign minister Moratinos joined his French counterpart, Michel Barnier, in exploring the possibility that certain privileged Mediterranean partners might be given access to EU structural funds. They proposed a pilot project in northern Morocco (Claret, 2005). Although the idea remained in the air (partly because the Western Sahara conflict got in the way of Neighbourhood CBC projects proposed by Spanish and Moroccan authorities), it is clear that some momentum was building up around bilateral co-operation in 2004–05, facilitated by changes in Spain's position on the Western Sahara, which under Zapatero became more supportive of Moroccan proposals. There was increased co-operation over migration and against Euskadi Ta Askatasuna (ETA), indicating a wider remit.

Spain thus seemed to be occupying a rather ambivalent position at this time in terms of the debate initiated by Christopher Hill (2004) about a 'renationalizing' trend in European politics. On the one hand, confirmation of the trend may be detected in the way that regional aspirations were lowered in the Mediterranean as projects with a 'global' or even sub-regional dimension were downplayed and a more modest set of national priorities concentrated upon. Spain warmed to the bilateralism of the ENP, seeing it as a vehicle of great value for its prioritized relationships in North Africa. On the other hand, Zapatero was keen to return to the European mainstream after withdrawal from Iraq, and especially to overcome Spain's alienation from France and Germany during the conflict. Moreover, the renationalizing course was a less viable path for small and medium-sized European states than for the larger ones if they wished to maintain international influence. The EU was still providing Spain with valuable structural funding and offered vital reinforcement in response to security challenges emanating from around the Mediterranean. A further reason for backing 'more Europe' in the field of CFSP was that Zapatero's immediate programme was primarily concerned with domestic reform. Finally, the new prime minister's ideological outlook may be germane to this issue. His early discourse as prime minister was replete with cosmopolitan values and belief in multilateralism, reflected in his 'Alliance of Civilizations' initiative; though not always a guide to his behaviour in office, such ideas may have found some reflection in the way Spain responded to Sarkozy's *démarche*: initially with silence and eventually in a constructive and co-operative tone, seeking compromise within the EU.

## Reacting to French Unilateralism

*Phase I (UM)*

In February 2007, Nicolas Sarkozy's initial call for the creation of an entirely new 'Mediterranean Union' took Spain completely by surprise.[8] There had been no prior consultation either bilaterally or through the Euro-Mediterranean framework, for the origins of the idea were linked to Sarkozy's ascent to the French Presidency, rather than representing an initiative by the Quai d'Orsay. Moreover, its immediate electoral purpose had more to do with domestic politics and Turkey's status vis-à-vis the EU than with the shortcomings of the EMP.

Up to this point, the multilateral aspects of French Mediterranean Policy had been developed primarily through the EU. To change the parameters and construct a union of littoral states was a radical (though hardly novel) idea, unwelcome to an increasingly Europeanized Spain. For the proposal implied a complete dismissal of the EMP, with which Spain was so intimately associated through the soubriquet of the 'Barcelona Process' and through several of the Partnership's major outcomes being associated with Spanish entrepreneurship. Moreover, ever since Felipe González had sold the idea of enhanced EU involvement in the Mediterranean to Helmut Kohl back in 1993, Spain had valued the involvement of non-Mediterranean member states in addressing security challenges emanating from the South. Besides liaising regularly with fellow Mediterranean member states, Spain had become habituated to collaborative activity that involved other EU actors. It was openly appreciative of the input of the European Commission and the Euro-Med outcomes of some of the northern EU Presidencies, such as those of Sweden in 2001 and Finland in 2006.[9] Spain had put years of effort into persuading northern member states that the Mediterranean was a *European* and not simply a southern European issue, and wanted to see further enhancement of the EU Mediterranean Policy rather than an alternative approach to the area. Thus it was deemed a serious mistake for France now to propose to dispense with the commitment of the EU as a whole, whatever its shortcomings when measured against the challenges.

At the same time, Spanish representatives did not feel they could openly oppose the UM formula of the French president. On the one hand, Spain's interest in working in close co-operation with France went far beyond the Mediterranean to issues such as the challenge from Basque separatist and Islamist terrorism, the management of migration flows and major infrastructural projects (energy networks, high-speed train links) that were important to the country's security and economic progress. The Mediterranean, ultimately, was not a decisive factor in the bilateral relationship. On the other hand, Spanish representatives still saw a need for southern member states to work together if the general cause of the Mediterranean were to be advanced within the EU. Thus, despite fundamental misgivings[10] and a lack of direct bilateral explanation of Sarkozy's initiative until October 2007, Spain at first maintained an official silence over the UM proposal. Collaborating closely with France, bilaterally and within the EU, was part of Zapatero's international strategy. Besides, it was hard for Spain to express criticism when it was privately celebrating the fact that, for the first time in years, France was again undertaking an ambitious initiative in the Mediterranean,[11] judging by all the talk of *grands projets* and new institutions and the flurry of seminars and workshops being held north of the Pyrenees. Sarkozy had focused European attention on the Mediterranean and had thus created an opportunity of sorts. Spain did not wish to discourage the dynamic Sarkozy from prioritizing the Mediterranean in his foreign policy, especially now that the eastern enlargement had increased the EU's focus on the East. Spain itself had been advocating EMP institutional reinforcement for some time and hoped that some impetus in this direction might be achieved if the Sarkozy initiative could be re-channelled towards the existing geographical parameters of the Partnership.

Rather than criticize the UM concept, Spain's foreign minister simply flew a different kite. Moratinos used the occasion of a university degree ceremony in Malta and a newspaper article to propose transforming the EMP into a 'Euro-Mediterranean Union' (*El País*, 2 August 2007). In this decidedly informal way, he outlined ideas for a new institutional architecture designed to upgrade political collaboration to the level of heads of state and government and to reinforce the parliamentary dimension. Significantly, however, there was no diplomatic campaign in support of his proposals. Spain was already involved in an EU working group in Brussels looking at the institutional enhancement of the EMP[12] and did not want to be seen to be initiating a *counter*-campaign against France. A final reason for Spain's delayed reaction is that, during the early months, the plans and statements emanating from France were inconsistent concerning the implications of the UM for the EMP.

Only after France had provided Spain with a presentation of the Sarkozy plan, through a visit by Ambassador Alain Le Roy to Madrid on 1 October 2007, did Spain begin to focus on working for a compromise between the French proposals and the existing realities surrounding the EU's central involvement in the EMP. Finding that

Romano Prodi shared his own reservations about the UM, Zapatero worked with his Italian counterpart in an attempt to persuade Sarkozy to modify the initiative.

The Spanish and Italian prime ministers thought they had achieved this through the *Appel de Rome*, issued at the end of a tripartite summit meeting with Sarkozy in December 2007.[13] Indeed, collaboration seemed to be ensuing from the compromise when Spain and France announced the creation of direct channels to liaise over the revised proposal to be presented to the EU for a 'Union for the Mediterranean'. None the less, the substance of the French formulation remained vague during the early months of 2008 and Germany, for one, was far from convinced that the Spanish–Italian manoeuvre had actually brought effective Europeanization of the original French proposal.[14] Germany saw it as evidence of weakness that Spain had not done more to defend the Barcelona Process before December (Schumacher, 2008: 8–9) and was annoyed that Angela Merkel had to confront Sarkozy herself at their meeting in Hanover on 3 March 2008, in order to secure an apparently unequivocal dilution of the initiative, leaving the EMP seemingly intact but with new institutional arrangements and a number of new high-profile projects.

*Phase II (UfM)*

France's determination to lead the new phase in Euro-Mediterranean relations, implicit in its decision not to present a detailed proposal to the European Council, only gradually became fully evident during the second half of 2008, during the French Presidency of the EU. It was seen in a variety of ways in the months following the Paris summit: (i) through bilateral negotiations with forthcoming EU presidencies (Czech, Swedish and Spanish) aimed at ensuring that France would remain UfM co-president for two years: that is, beyond its term as EU president, until 2010; (ii) through resisting suggestions that EU common positions should be adopted on the UfM (though they were used in working group discussions on the structure of the new Secretariat); and (iii) through Sarkozy's active Middle East diplomacy, which continued even after he ceased presiding over the European Council at the end of December 2008, complicating France's promotion of the UfM. Symbolically, the French ambition to lead was underlined by its successful move to have the 'Barcelona Process' prefix removed from the name of the UfM in November 2008, less than six months after it had been adopted in accordance with a European Council decision of March 2008.[15] These indications suggested that France, even now that the parameters of the UfM had become coterminous with those of the EMP, was still using unilateral initiative and bilateral intergovernmental bargaining as means of minimizing northern European influence, in the interests of French protagonism at the Mediterranean (and ultimately European) level.

Thus, Spain quickly discovered that in fact it had not safeguarded the future of the approach to EU Mediterranean Policy associated with the Barcelona Process. It might have been expected now to adopt a more questioning attitude. After all, EU concern over French Mediterranean policy was widespread and there was a built-in opportunity for Spain to stand on principle by insisting on co-presiding (with Egypt) over the UfM during its own EU Presidency. Spain's decision not to distance itself from France even now was seen by conservative opposition at home as a sign of government weakness, yet it reflected not only Zapatero's decision but also the view of diplomats that working with France would be a better option than any alternative.[16] The same thinking that had dissuaded Spain from rejecting the earlier UM proposal was still present, shaping Spanish reactions to the unfolding reality of the UfM, although now it was encouraged by France's apparent readiness to allow Spain a significant role in the new venture.

France's wider importance for Spain, its renewed enthusiasm for the Mediterranean, the need for countries with special Mediterranean interests to work together, genuine identification with substantive aspects of the UfM and recognition of France's greater capacity to take forward the more ambitious agenda, which by now was supported by a lot of national preparation – all these considerations remained relevant to Spain's reticence to express divergent criteria. Even when France frustrated the widespread EU expectation that the European terms of participation in the UfM co-Presidency would coincide with the six-month rotating EU presidencies, Spain did not defend this principle. Rather, it declared that it would follow the established pattern, whatever that was, thus putting the onus on Sweden, a smaller country, to either stand up for its EU right to preside over external relations during its (preceding) Presidency or to yield to French pressure to concede an agreement similar to that already reached with the Czech Republic. Playing down the issue, Spanish diplomats expressed hopes that ratification of the Lisbon Treaty would finally resolve this matter.

More self-interested motives for waiving the right to co-preside over the UfM during its EU Presidency also account for Spain's position. A bilateral 'deal' was done, with France in November 2008 supporting the candidature of Barcelona to host the permanent Secretariat in return for Spain accepting that 'Barcelona Process' would disappear from the name of the UfM (Barbé, 2009: 39).[17] Spain's acceptance of the pattern of two-year European terms in the UfM Presidency, for which practical justification was claimed in the potential benefits to be gained from accumulated experience, was also informed by the Spanish interest in presiding for a two-year period itself, in 2010–12, if the Lisbon Treaty were not ratified or responsibility delegated by the European Council. France was prepared to support Spain over this, although whether it could deliver was another matter, in view of reservations expressed by northern member states.

At this time, Spain attached prime importance to securing the Secretariat, against the rival candidature of Malta. Hosting the first genuinely Euro-Mediterranean political institution would be a way of linking the UfM with the tradition of the Barcelona Process and making the city a headquarters for Euro-Mediterranean collaboration. In contrast to intensive lobbying to win this contest, Spain showed little interest in the political positions (of secretary-general and deputy secretaries-general) that were available within it, nor did it emerge immediately as an instigator of high-profile projects that it might aspire to be at the centre of.[18] It may be that national ambition regarding projects was curbed by the particular severity with which the global recession hit Spain, but the criticism that the government was more preoccupied with branding than with substance (Kausch, 2010: 5) was not exactly a new one.[19]

Eventual multilateral agreement on the statutes of the new Secretariat and its inauguration in March 2010 gave the impression of an emerging partnership with France, facilitated by a close working relationship between the then foreign ministers, Bernard Kouchner and Miguel Ángel Moratinos. While evidently Spain had been more comfortable with the European-managed EMP that existed until 2008, it was still able to trade on the political capital it had derived from the Barcelona Process and potentially stood to benefit from France's need to overcome the acute EU isolation caused by its Mediterranean diplomacy during the initial UM phase. The need to liaise over Secretariat negotiations and the Mediterranean component of the Spanish EU Presidency intensified contact between France and Spain over the UfM. With the UfM still not 'consolidated',[20] the two countries needed each other and could not afford a disagreement.

This begs the question: how much consensus exists in French and Spanish outlooks on the Mediterranean? Will points of divergence and latent historical rivalry make this an unlikely or uneasy marriage of convenience, or is there enough underlying unity for French–Spanish co-operation to become effective, even to the point of creating synergy? Limited thus far, an increase in collaboration, while dependent on greater commitment by Spain, might be one scenario to provide the UfM with the 'motor' it has lacked during its attempted takeoff.

## Spanish and French Positions Compared

Spanish and French representatives have acknowledged some areas of divergence in national perspectives on Mediterranean policy.[21] Based on interviews conducted with French and Spanish personnel involved in the UfM, convergence and divergence between the two countries can be summarized as follows:

### *(i) Status of UfM in Relation to the EU and its Approach to Euro-Mediterranean Relations*

While Spain accepted the French push towards intergovernmental dealings in Euro-Mediterranean relations, this is not something it would have encouraged itself. As a 'smaller' member state, it is a more natural team player and has looked to entrepreneurship and alliances rather than a unilateral style of leadership in order to exert influence in Euro-Mediterranean relations. With the exception of the ill-fated attempt by José-María Aznar to ally with Britain (and the US) in the run-up to the Iraq War, Spain has not been part of an informal 'axis' trying to shape EU policy and indeed has been a stalwart of the CFSP. From the start, unlike France under Sarkozy, it has proposed a strong, engaged role for the European Commission in the UfM.

### *(ii) UfM Relationship to EMP*

While France has been inconsistent over whether the UfM should enhance or replace the EMP and has said relatively little about Euro-Mediterranean activity beyond the prioritized UfM projects, Spain emphasizes enhancement, not only to protect its own legacy but also because it values the broader set of objectives provided by the Barcelona Declaration. This difference should not be exaggerated, however, since both countries lack enthusiasm for direct Euro-Mediterranean activity to promote democracy and human rights (Kausch, 2010: 3).

*(iii) UfM and ENP*

While neither country has been explicit about the relationship of the UfM to the ENP, France tends to see the latter as a 'German-friendly concept', of little relevance to the Mediterranean (Schmid, 2009: 73), whereas Spain sees it as a valuable framework through which to accommodate differentiated North–South relations at the bilateral level and has tried to use its sub-regional CBC programme as well. Spanish diplomats value the proven stability of relations within the ENP framework and feel that it offers scope for reinforcing relations with willing partners (Morocco, Tunisia, Egypt) even if some neighbours have opted out. Unlike the UfM, the ENP is not going to stall, whatever the vicissitudes of relations with individual EU partners.

*(iv) Mediterranean Vision and Leadership*

While current French Mediterranean policy is informed by a 'vision', albeit a very old-fashioned one (Rubio Plo, 2008), Spanish policy is more concerned with interests and institutions and has been criticized for 'narrow concerns' and a lack of vision (Kausch, 2010: 1). Colonial histories have left both countries claiming a special understanding of Mediterranean issues and each sees the littoral states constituting a region, yet French policy is distinguished by Gaullist ideas that make *unique* claims to regional leadership as an extension of the French notion of a national civilizing mission. Hence Sarkozy's lauding of France's colonial involvement during visits to North Africa in 2007. The French president aspires to leadership of the Mediterranean 'region' and influence in Middle East peace initiatives. Zapatero is more of a multilateralist, as seen in the uploading of his Alliance for Civilizations project to the UN and his willingness for Herman Van Rompuy to preside over summits during the Spanish EU Presidency, not to mention the more modest, discreet style normally associated with Spain's Middle East peace diplomacy.

*(v) Region-building*

While both France and Spain have demonstrated their interest in region-building and have special interests in the Maghreb, only France has pushed for an exclusively Mediterranean union. Both countries now work within the parameters of the UfM, but they disagree over the desirability of non-Mediterranean EU countries playing prominent roles (e.g. through the co-Presidency), France dismissing the idea. And while Sarkozy has used functionalist arguments to present major UfM projects as the new foundations for region-building, Spain has a more multi-faceted notion of region-building associated with the Barcelona Process (albeit with greater scepticism since the 1990s).

*(vi) Institutions*

Both countries have been keen to see Euro-Mediterranean relations become more institutionalized and involve regular meetings at the highest political levels, but in private Spanish diplomats express reservations about the viability of the co-Presidency and the commitment to hold summits every two years.[22] At times, each has seemed keen on the idea of a strong Secretariat with the potential eventually to extend its mandate into traditional areas of EMP activity, but Spain alone has argued for high-level European Commission involvement[23] and general compatibility with EU mechanisms; it has seemed less keen than France on having a powerful secretary-general in charge of the Secretariat.

*(vii) Co-ownership*

Both France and Spain have been committed to the idea of co-ownership, at least to the extent of giving southern partners positions in the new UfM institutions and formalizing the involvement of the Arab League, defended as a modest move towards North–South rebalancing.

*(viii) Projects*

While Spain has welcomed the French idea of large-scale technical projects, so far it has not come up with much in terms of its own proposals, this being in part a reflection on the far lower number of Spanish officials involved with the UfM and with development work in North Africa. As noted already, Spain has become involved in the Solar Energy plan (led by France and Germany) and – in collaboration with Italy – has proposed the creation of an agency to promote the development of small and medium-sized enterprises.

*(ix) Civil Society*

Although each country's discourse emphasizes the role of civil society organizations in the building of Euro-Mediterranean relations, this was not reflected in French institutional designs for the UM/UfM, nor has there been a Spanish enhancement initiative in this direction. None the less, a substantial programme of civil society activity, more thematically structured than in the past, was planned by NGOs to coincide with the Spanish EU Presidency and the opportunity to involve civil society actors may grow once the Secretariat is fully functional.

Overall, one finds here sufficient agreement to support the prospect of ongoing co-operation between France and Spain, yet evidence too of scope for Spain to impress a distinct personality on the UfM, especially if it were to play a future role in the European component of the co-Presidency. Spain may well continue to be overshadowed by France in the new enterprise, and there is little prospect of it becoming a paradigm shifter, but there are signs that it will resist any attempt at actual mentoring by France (*El País*, 29 April 2009). A lot may depend on who fills the key Mediterranean posts in the new European External Action Service (EEAS).

## A 'Unique Opportunity' Lost?

Spain's fourth EU Presidency (January–June 2010) took place in much more constraining circumstances than its previous two, in 1995 and 2002, both noted for Euro-Mediterranean innovation. Besides the context of economic recession, the Presidency commenced in tandem with the new positions of president of the European Council and high representative for foreign affairs and security policy established under the Lisbon Treaty. Moreover, rather than start with a clean slate, Spain inherited an unwelcome legacy of postponed UfM ministerial conferences, mostly as a result of the vulnerability of the new institutional structures to fallout from the Israeli–Palestinian conflict, especially following the attack on Gaza. Of 16 such conferences scheduled for 2009, no fewer than nine had been postponed while the cancellation of the Euro-Mediterranean foreign ministers' conference, due to be held in Istanbul in November, really underlined the extent of the malaise (Martín, 2009: 5). The Spanish Presidency wanted most of these conferences to take place eventually, but decided to hold only two of them in Spain;[24] its attitude was 'better fewer, but better', meaning that those it *did* take direct responsibility for (conferences on water management and tourism) should be meticulously prepared, productive and well followed up, and not so many in number as to detract Spanish officials from devoting sufficient time to higher profile events such as the first EU–Morocco summit meeting, held in Granada in March, and the second UfM summit, scheduled for 7 June in Barcelona.

Conscious that the first rotating Presidency under the Lisbon Treaty would set a precedent for subsequent ones, Zapatero announced beforehand that his main objective was to ensure that the new EU leadership figures gained visibility and became forces for cohesion (*El País*, 16 December 2009). While he could have tried to invoke a transition period in an attempt to retain more of the traditional role of the rotating president, Zapatero made clear that he wanted Herman Van Rompuy to preside over the substantial number of bilateral summits that were envisaged (14) and not simply meetings of the European Council. Spain, however, would be the main organizer of the summits, and thus be in a position to exert some influence. In terms of geographical focus, the Spanish Presidency aimed to be, first and foremost, a 'Euro-American Presidency', committed to developments in relations with the US, Canada, Latin America and the Caribbean, but it was also presented as a 'Euro-Mediterranean Presidency', aiming to strengthen EU relations with the Maghreb countries in particular.

Following the massive disruption to UfM meetings in 2008–09, one of the essential concerns of the Spanish Presidency was that decisions that had been taken already should be fully implemented, through having the Secretariat functional by the time of the envisaged summit in Barcelona,[25] by resolving lingering uncertainties about the modalities and renewal of the co-Presidency and by ensuring that the new UfM projects all migrated finally from the drawing board to implementation. In conjunction with the co-Presidency, a lot of diplomatic effort was expended in seeking a satisfactory turnout for the summit and on working for consensus around final UfM design details and a Work Programme for the next two years. The summit was viewed as crucial to the consolidation of the UfM. However, although the Secretariat was inaugurated during the Spanish Presidency, it was not able to start work immediately owing to disagreements over its funding arrangements; and, ultimately, the effort to demonstrate progress fell flat when a decision to postpone the summit (in principle until November) was finally announced in May.

The budgetary issue provided evidence of ongoing internal EU misgivings over the UfM while the postponement of the summit was a result of the limited progress of the initiative and its continuing contamination by fallout from Middle Eastern conflicts. Despite Spanish efforts to come up with a formula to allow for Israeli participation while avoiding the presence of foreign minister Avigdor Lieberman (a highly controversial figure with whom Egyptian and other Arab representatives simply refused to sit down), in the end the event was deferred rather than risk an Arab boycott or a deadlock at the summit. Earlier, in April, Middle East sensitivities had led to a failure to adopt an unprecedented Euro-Mediterranean strategy for the management of water resources, for although a draft document was produced, Israel's refusal to accept references to the Palestinian 'Occupied Territories' meant that it could not be adopted.[26] As in the institutional structures, political discord was intervening in the technical areas of UfM activity, frustrating the functionalist logic of Sarkozy's original proposal.

The only successful Euro-Mediterranean event with public resonance during the Spanish Presidency was the EU–Morocco summit. This went ahead largely as planned, confirming the recent upgrade in bilateral relations without defining its significance in more substantive terms. It brought home to Spanish representatives what 'life under Lisbon' really meant, for not only did Van Rompuy assert himself in his new role: the restructuring of the EU external relations hierarchy left foreign minister Moratinos entirely sidelined.[27] With the role of the rotating presidency reduced, Spain (along with other member states) will need to rely on placing senior officials in key posts within the EEAS or on external relations responsibilities being delegated from the high representative to Spaniards in order to exert strong national influence within the UfM framework. Of course, there are still opportunities through the Secretariat, to which countries may second officials, but so far Spain has been preoccupied with merely hosting this institution rather than filling senior positions.

The prospect of a successful relaunch of the UfM through a second summit during the Spanish Presidency was thus dashed, leaving an unimpressive balance-sheet insofar as the Mediterranean was concerned. To see this as a lost opportunity is only partially fair, however. The opportunity for an influential Spanish contribution to the UfM at this time was clearly circumscribed by the entry into force of the Lisbon Treaty and the lack of progress in Middle East peace diplomacy. Spain – through its foreign minister – had done what it could to facilitate peace talks and to promote a more pro-active EU role, but was criticized by some actors[28] for trying too hard to use the UfM to these ends, rather than emphasizing a pragmatic agenda for the high-profile meetings. Earlier, Spain had toyed with the idea of trying to explore peace initiatives collectively with the Arab countries, through holding a conference in Cairo, but the idea was not pursued, presumably out of concern that it would be misinterpreted by Israel, the US and some fellow EU member states.[29] In August 2009, Sarkozy again surprised Spanish representatives by speaking of the possibility of holding a peace summit in Paris involving the Israeli and Palestinian leaders together with the UfM co-Presidency if conditions improved in the Middle East. Though widely seen as an ill-judged initiative, Spain did not distance itself from this move, which seemed to link the future of the UfM more than ever to prospects of Middle East peace diplomacy: indeed, Zapatero himself was now saying that the postponed UfM second summit would be arranged when there was some certainty that it would be a 'positive factor vis-à-vis the Middle East situation' (*El País*, 8 September 2010).

In other respects, the Spanish EU Presidency definitely *was* a lost opportunity. Certainly, the government's scope for initiative was reduced dramatically during this period, with pressure from the financial markets, the EU and the USA forcing Zapatero to announce more radical austerity measures in May. Yet earlier, before international concerns arose that other southern European countries might suffer the same fate as Greece, some Spanish observers were pointing to a 'unique' opportunity that clearly was not being taken on the eve of the Presidency (Martín, 2009: 1): the argument was that even a modest mobilization of additional diplomatic and financial resources might have made a difference to the ideational and organizational aspects of the UfM's development during its EU Presidency. While some saw the country's administrative capacity as adequate to the immediate task (Soler i Lecha and Vaquer i Fanès, 2010: 74–5), others made discomfiting comparisons between the sizes of the Spanish and French UfM teams (Martín, 2009: 7).[30] The diminutive task force can be related to a lack of clarity concerning Spain's strategic priorities and possibly a trimming of ambition, the latter reflected in the decision to host only two UfM ministerial conferences, and to choose topics that, despite the importance of water and tourism, hardly seemed likely to arouse popular enthusiasm for the UfM.

The opportunity may have been 'unique' because of the changing architecture of EU external relations and the fact that Spain was by no means assured of a mandate to fill the role of European co-president of the UfM for the next two years, although some analysts assumed that this would be so (Aliboni, 2009: 6). Moreover, it was a rare opportunity for the UfM itself, given that the Spanish EU Presidency followed two rotating presidencies (Czech Republic and Sweden) lacking in Euro-Mediterranean distinction and was being followed by other presidencies (starting with Belgium) of which there were no great expectations in this regard.

Given Spain's record in relation to the Barcelona Process, it was not unreasonable to expect that its Presidency would prove useful, at least at the modest level of getting the new UfM institutions working with some semblance of normality, thus laying the ground for practical progress at the level of projects. If this could be achieved, there was some possibility of reassuring EU actors and their partners that the UfM venture was still viable and worth persisting with. However, Spain was not able to make a difference during its EU Presidency, thus leaving doubts about it providing effective political orientation for the UfM in the future. While scepticism arose partly from the general shortcomings of Zapatero as Spain's chief foreign policy actor (de Areilza and Torreblanco, 2009; Grant, 2009), it was in no way diminished by the evident lack of interest in the Mediterranean within the opposition Partido Popular, which by 2010 was enjoying a substantial lead over the Socialists in opinion polls.

## Conclusion

The misfortunes that befell the Euro-Mediterranean agenda of the Spanish Presidency leave much uncertainty about Spain's future as a driving force within the UfM. In 2010 Spain did manage to recover the visibility it had lost in 2007–08 amid the flurry of French proposals (Barbé and Soler i Lecha, 2009: 97). It returned to the stage once more, though this time – owing to European-level developments – accompanied at key events by Herman Van Rompuy and also, throughout, by France, as the European co-president of the UfM. With some inevitability, Spanish leaders were overshadowed by EU representatives at the summit with Morocco in Granada and would have been so again at the UfM summit in Barcelona if it had taken place. Rather more damaging to Spanish prestige were the non-events: the 'postponements' that were announced or the meetings that ended in failure, notably the water conference, whose failure sent truly negative signals about the seriousness of the UfM.

None the less, Spain has acquired an embedded role in the UfM through its hosting of the Secretariat. This is certainly an achievement, although ambiguities in its statutes leave open the question of just how central this body will be to future Euro-Mediterranean relations. If Spain were given an additional role in the UfM co-Presidency, this would also add to its prominence in the short term, while in the longer term the placing of Spanish officials in key positions in the EEAS will affect national influence. However, it is not clear what a Spanish UfM co-president would be able to achieve, since Spain's successors in the rotating EU Presidency (Belgium and Hungary) are not keen for Spain to continue chairing ministerial conferences.[31] Having lost powers under the Lisbon Treaty, rotating EU presidencies may become more determined to hang on to their remaining prerogatives, not least in respect of the UfM. For Spain there is thus some risk of raising expectations that cannot be fulfilled because even a term in the co-Presidency, without adequate domestic backup and Euro-Mediterranean support, might bring rather modest levels of influence in reality (especially when the role of European co-Presidency can only be performed in conjunction with the southern co-Presidency). Visibility, a presence on the stage, could become disadvantageous.

Even if Spain manages to consolidate its prominence at the institutional level, France can be expected to remain influential as well, by exploiting the intergovernmental aspects of the UfM, landing key positions in the EEAS and investing (as it already does) unparalleled political energy and resources in the Union for the Mediterranean. It will continue to take initiatives and centre a considerable amount of activity in France. While a clarification of national strategic priorities is certainly a precondition for Spain to regain influence in the Euro-Mediterranean context, in itself this would not be enough to overturn French dominance of the UfM so long as the Elysée continues to regard the UfM as a priority, to be pursued vigorously.

Given the relative standing of France and Spain within the EU and internationally and the formidable economic challenges it faces, Spain is not going to suddenly start competing with France for influence in the foreseeable future – especially if international confidence in the southern European economies were to decline further. Spain clearly has limitations then as a hypothetical 'leader', yet it might be able to turn its apparent shortcomings to some advantage. Spain may be able to outperform France, if only in terms of acting as a 'broker' or 'entrepreneur', roles very much needed within the EU and UfM alike. At the end of the day, as Spain recognized early on, the two countries need to pull together, and be capable of promoting wider collaboration, if the UfM is to fulfil its objectives, though there are limits to what even a Franco-Spanish alliance could achieve in the enlarged EU.[32]

At the present time, co-leadership is at the most incipient, indeed little more than a spectre. It continues to meet resistance within France, while also generating objections from some northern and eastern EU member states. The conundrum for Spain is whether it can join forces with France to provide the stronger 'motor' that a diminished Mediterranean lobby seems to need in the context of the enlarged EU, *without* alienating from the UfM the majority of states that lack strong Mediterranean interests. By 2010, many member states were already sending more junior representatives to senior officials' meetings. The breadth of EU commitment on show at the founding Paris summit may erode further if influential European actors decide that the UfM is dominated by a clique, especially if projects such as the Solar Plan appear to give a competitive advantage to companies based in 'insider' countries. This would represent a failure of policy entrepreneurship, leaving France and Spain in the forefront (in the Spanish case, more institutionally than strategically) in an enterprise going nowhere.

## Acknowledgements

Research for this paper included a series of interviews with Spanish diplomats, policy advisers and think tank experts and EU officials, conducted in 2009–10. It was made possible by an award from the British Academy for a project on 'The Union for the Mediterranean: Significance for the Barcelona Process' (SG-51979). I am grateful to Elisabeth Johansson-Nogués, Federica Bicchi and Iván Martín for comments on an earlier draft.

## Notes

[1] For example, Sweden's eventual acceptance of France remaining in the UfM co-Presidency during its own EU Presidency in 2009 has been widely related to French support for the Eastern Partnership promoted by Sweden.

2 Countries mentioned to the author by an EU official in March 2010 included the UK, Latvia, Finland and Sweden.
3 Various factors led to partial moderation: France's need to negotiate with successive rotating EU presidencies; the dynamics of the working group that drafted the UfM Secretariat statutes in 2008–10, since this proceeded by consensus; and the desire to draw on Commission expertise and resources. Interview with a European Commission official, March 2010.
4 Aliboni (2009: 8) none the less refers to 'the emerging France-Spain duo'.
5 However, the European Commission representative in the Secretariat, Andrés Bassols, is Spanish.
6 The failed attempt in 1996–2000 to adopt a Charter on Peace and Stability was the most crucial tangible manifestation of this trend.
7 Interview with a Spanish diplomat, May 2009.
8 Interviews with Spanish diplomats, May 2009.
9 Interviews with Spanish diplomats, May 2009.
10 Soler (2008: 5) sees Spain's initial reaction as possibly the most negative among the Latin countries, fearing that the Barcelona Process would be damaged and Spain's role in the Mediterranean eclipsed. While most Spanish representatives showed their coolness through silence, unrestrained condemnation of the UM as 'a unilateral and self-interested attempt to torpedo the Barcelona Process' and to 'take back the European baton in the Mediterranean' was expressed by former Spanish Euro-Mediterranean ambassador Juan Prat, by this time ambassador to the Netherlands. Prat went on to claim that, through the UfM, the initiative had ended in a move to consolidate the Barcelona Process (Prat y Coll, 2008: 46, 48).
11 Chirac's second Presidency had emphasized the *politique arabe* rather than a Mediterranean policy (Schmid, 2009: 73).
12 This pre-dated Sarkozy's initiative.
13 'Appel de Rome de la France, de l'Italie et de l'Espagne', 20 December 2007, included in a French 'Note d'étape au 19 janvier 2008' (mimeo). The document stated that the UfM would complement and co-operate with existing institutions involved in Mediterranean co-operation and would not affect the processes of stabilization or association of the countries concerned, or the accession negotiations of Croatia or Turkey.
14 Barbé and Soler i Lecha (2009: 93) describe Spain's response to the UM proposal as 'soft Europeanization', contrasting with Germany's 'hard Europeanization'.
15 Earlier, Spain had lobbied to get its Barcelona copyright mentioned in the name of the UfM in order to signal that the EMP was not, finally, being replaced (Soler, 2008: 25). Moratinos rationalised its eventual elimination by saying the EMP was now a true 'union' and no longer involved a mere 'process' (Hernando de Larramendi, 2009: 52).
16 Personal interviews with Spanish diplomats, May 2009.
17 This was confirmed to me by a French diplomat interviewed in April 2009.
18 Spain's only input here was through a modest Mediterranean Initiative for Entrepreneurial Development, involving support for small and medium-sized enterprises; planned with Italy since 2006, this was now adopted as a UfM project. There was also Spanish interest in participating in the Solar Plan, developed by France and Germany (Moratinos, 2009: 18).
19 Sharing this view, the *Financial Times* ('A Stumbling Spain', 6 January 2010) described the work programme of the Spanish EU Presidency as 'remarkably anodyne' and criticized its emphasis on the Lisbon Treaty for privileging institutional arrangements over the problems affecting European citizens.
20 As recognized by Moratinos and Kouchner in a joint article with Egyptian foreign minister Ahmed Aboul Gheit, 'Comprometidos por el Mediterráneo' (*El País*, 4 March 2010).
21 One Spanish Foreign Ministry official put the level of agreement between the two countries at '95 per cent' (personal interview, Madrid, 8 May 2009).
22 See Fundación Alternativas/Friedrich Ebert Stiftung (2008), for the proceedings of a seminar involving French and Spanish diplomats: French ambassador Jacques Huntzinger described the value added by the UfM as residing in its new political institutions while his Spanish counterpart Fidel Sendagorta maintained that strong institutions could not be built while there was instability in the Middle East.

Several Spanish interviewees expressed doubts about the decision to institutionalize summits, given the risks of cancellation, poor attendance or disunity. Interestingly, they did not mention any concrete political objectives to be advanced through a summit, the attitude of one respondent being that, 'at the end of the day, a summit is a photo'.

[23] See, for example, *ENPI Information* (Brussels), 4 March 2010.

[24] The others were planned to take place in Dubrovnik, Cairo and Brdo (Slovenia).

[25] Barcelona became an obvious venue for the summit once it became clear that the co-ownership principle of holding summits alternatively in the North and South remained unviable in the absence of an Israeli–Palestinian peace settlement.

[26] *ENPI Information*, 27 April 2010.

[27] The press conference following the Granada summit was presided over by Van Rompuy, not Zapatero. This made a difference in that he called on Morocco to improve its human rights situation (*El País*, 8 March 2010).

[28] Interview with an EU official, May 2010.

[29] Spain hosted a conference with Arab foreign ministers in January 2007 and spoke of following it up with a meeting at prime ministerial level. The idea of co-organizing an informal conference with Egypt in Cairo to generate movement towards wider Middle East peace efforts during Spain's EU Presidency was announced by deputy PM María Teresa Fernández de la Vega after a meeting with President Mubarak in April 2009 (*El País*, 27 April 2009), but nothing came of it.

[30] According to Martín, the French UfM mission has consisted of between 12 and 20 civil servants and experts and there is multi-ministerial involvement, whereas Spain had only an ambassador on a special mission with support from another diplomat and three temporary advisors (for the EU Presidency Mediterranean activity), based in the Foreign Ministry.

[31] Interview with an EU official, March 2010.

[32] For example, their joint opposition could not prevent the European Commission from adopting new ENP financial allocations for 2011–13 that were more favourable to the eastern than the southern neighbourhood.

## References

Aliboni, R. (2009) The Union for the Mediterranean: evolution and prospects (Rome: Istituto Affari Internazionali, IAI, Documenti IAI 0939E).

Barbé, E. (2009) La Unión por el Mediterráneo: de la europeización de la política exterior a la descomunitarización de la política mediterránea, *Revista de Derecho Comunitario Europeo*, 13(32), pp. 11–48.

Barbé, E. & Soler i Lecha, E. (2009) What role for Spain in the Union for the Mediterranean: Europeanising through continuity and adaptation, *Hellenic Studies*, 17(2), pp. 85–102.

Bicchi, F. (2006) The European origins of Euro-Mediterranean practices, in: E. Adler et al. (Eds) *The Convergence of Civilizations: Constructing a Mediterranean Region* (Toronto: University of Toronto Press).

Bicchi, F. (2007) *European Foreign Policy Making toward the Mediterranean* (New York: Palgrave).

Cembrero, I. (2006) *Vecinos alejados: Los secretos de la crisis entre España y Marruecos* (Barcelona: Galaxia Gutenberg/Círculo de Lectores).

Claret, A. (2005) France and Spain Bring Positions Together in the Maghreb, *The Mediterranean Yearbook: Med.2005* (Barcelona: Institut Europeu de la Mediterrània).

De Areilza, J. M. & Torreblanca, J. I. (2009) Diagnóstico diferencial, política exterior, *Foreign Policy*, 33 (Madrid: Spanish edition). Available at http://www.fpes.org (accessed 2 June 2009).

Fundación Alternativas/Friedrich Ebert Stiftung (2008) 'La Unión para el Mediterráneo y el reforzamiento del núcleo euromediterráneo', seminar proceedings, Madrid, 23 May.

Gillespie, R. (1995) Spain and the Maghreb: towards a regional policy? in: R. Gillespie, F. Rodrigo & J. Story (Eds) *Democratic Spain: Reshaping External Relations in a Changing World* (London: Routledge).

Gillespie, R. (2000) *Spain and the Mediterranean: Developing a European Policy towards the South* (Basingstoke: Macmillan).
Gillespie, R. (2002) The Valencia Conference: reinvigorating the Barcelona Process? *Mediterranean Politics*, 7(2), pp. 105–114.
Grant, C. (2009) *Will Spain Remain a Small Country?* (London: Centre for European Reform).
Hernando de Larramendi, M. (2009) The Mediterranean policy of Spain, in: I. Schäfer & J.-R. Henry (Eds) *Mediterranean Policies from Above and Below* (Baden-Baden: Nomos).
Hill, C. (2004) Renationalizing or regrouping? EU foreign policy since 11 September 2001, *Journal of Common Market Studies*, 42(1), pp. 143–163.
Kausch, K. (2010) Spain's diminished policy in the Mediterranean, FRIDE Policy Brief, no. 26 (Madrid: Fundación para las Relaciones Internacionales y el Diálogo Exterior).
Martín, I. (2009) Las prioridades de la Presidencia española de la UE en el Mediterráneo: ser y debe ser, ARI Paper, 166 (Madrid: Real Instituto Elcano). English translation available at http://www.realinstitutoelcano.org/wps/portal/rielcano_eng/Content?WCM_GLOBAL_CONTEXT=/elcano/elcano_in/zonas_in/ari34-201 (accessed 18 February 2010).
Moratinos, M. A. (2009) Unión por el Mediterráneo: un nuevo paradigma anclado en el acervo de Barcelona, *The Mediterranean Yearbook: Med.2009* (Barcelona: Institut Europeu de la Mediterrània).
Núñez Villaverde, J. A. (2005) Spanish Policy towards the Euro-Mediterranean Partnership, in: H. Amirah-Fernández & R. Youngs (Eds) *The Euro-Mediterranean Partnership: Assessing the First Decade* (Madrid: Real Instituto Elcano and Fundación para las Relaciones Internacionales y el Diálogo Exterior).
Prat y Coll, J. (2008) The European Project in the Mediterranean: Barcelona Plus, *The Mediterranean Yearbook: Med.2008* (Barcelona: Institut Europeu de la Mediterrània).
Rubio Plo, A. R. (2008) La política mediterránea de Francia: del imperio latino de Alexandre Kojève al neogaullismo de Henri Guaino, ARI Paper, 86 (Madrid: Real Instituto Elcano).
Schmid, D. (2009) French ambitions through the Union for the Mediterranean: changing the name or changing the game? *Hellenic Studies*, 17(2), pp. 67–84.
Schumacher, T. (2008) From Paris with love? Euro-Mediterranean dynamics in the light of French ambitions, *The GCC–EU Research Bulletin*, 10(May), pp. 8–10.
Soler, E. (2008) España y el Mediterráneo: En defensa del Proceso de Barcelona, in: E. Barbé (Ed.) *España en Europa 2004–2008*, Monografías del Observatorio de Política Exterior Europea, 4 (Barcelona: Institut Universitari d'Estudis Europeus).
Soler i Lecha, E. & Vaquer i Fanès, J. (2010) The Mediterranean in the EU's Spanish Presidency: a priority in turbulent times, *Mediterranean Politics*, 15(1), pp. 73–89.

# Germany and Central and Eastern European Countries: Laggards or Veto-Players?

## TOBIAS SCHUMACHER
ISCTE – University Institute of Lisbon, CIES-IUL, Lisbon, Portugal

ABSTRACT *Germany, Poland, Hungary and the Czech Republic have always been loyal members of the Euro-Mediterranean Partnership (EMP). Yet they have never shown enthusiasm for it or displayed noteworthy activism within a co-operation framework that is perceived by their foreign policy elite as a necessary concession to southern EU member states in order to secure the latter's support for continuing pro-active EU engagement in eastern Europe. With this in mind, this study provides a comparative analysis of the foreign policy of Germany, Poland, Hungary and the Czech Republic toward the southern Mediterranean and discusses in particular the different responses of their governing, economic and societal elites to plans to create a Mediterranean Union (UM).*

Germany and the three central and eastern European countries Poland, Hungary and the Czech Republic share the fact that the end of the Cold War had a tremendous impact on their foreign policy. Following the Second World War, Germany was a divided and non-sovereign state while Poland, Hungary and what used to be called Czechoslovakia became an integral part of the Soviet Union's sphere of influence. For decades they were on the front line in the East–West conflict – a situation that considerably restricted their foreign policy behaviour and room for manoeuvre. As a consequence, these four countries also shared and share the common characteristic that the southern Mediterranean hardly featured on their foreign policy agendas.

The end of the Cold War brought an end to the division of Germany and to Soviet-inspired Marxism-Leninism in central and eastern Europe and generated the structural preconditions for all four countries to regain full autonomy in foreign policy matters. While, for Germany, this has led to the adoption of a more global foreign policy agenda and further accentuation of its commitment to multilateralism, its western orientation and its Euro-centrism, Poland, Hungary and the

Czech Republic opted for a more regional outlook and a foreign policy that is primarily concerned with the integration of eastern European countries into the EU – in parallel to their full integration into transatlantic political and military structures. Yet, on account of their participation in the Barcelona Process and what now is called the Union for the Mediterranean (UfM), as well as the coming into force of the Schengen agreement and the creation of a common EU external border with the southern Mediterranean, all four countries have become much more exposed to Europe's southern neighbourhood.[1]

Against this backdrop this study aims to provide a comparative analysis of the extent to which the southern Mediterranean and Euro-Mediterranean relations feature on these four countries' individual post-Cold War foreign policy agendas. On the basis of this analysis it will discuss how the UfM and the process leading to its creation were perceived by the four countries' governing, economic and societal elites and will analyse how far these perceptions resulted in governmental action and/or policy entrepreneurship. Ideational factors, the importance given to a specific foreign policy matter, the cultural and historical legacies, the existence of potentially powerful domestic constituencies, and structural geo-strategic parameters condition the foreign policy decision-making process and, thus, governmental policy responses. Accordingly, these factors will be incorporated and underpin the analysis. The study is guided by the assumption that the UfM is unlikely to alter the pattern of the four countries' behaviour in Euro-Mediterranean settings and reinforces, rather than diminishes, their role as veto-players, favour-exchangers and unhappy laggards.

## Germany

*German Foreign Policy and the Mediterranean: Actors, Interests and Issues*

The fall of the Berlin Wall in November 1989 and Germany's subsequent unification process had a tremendous impact on its foreign policy, its international outlook and its perceptions of the southern Mediterranean. Naturally, as it was a divided and non-sovereign state that 'imported' its security from the United States, while simultaneously guaranteeing it through NATO membership, its autonomy in foreign policy matters was heavily constrained for more than four decades. As a result of this limited room for manoeuvre, German foreign policy confined itself for many years to issues 'very close to home' (Webber, 2001: 5). The southern Mediterranean was mainly perceived by the political elite from a developmental perspective and, within the larger NATO and hence Cold War context, from a strategic angle. While, for domestic economic actors, the (non-Arab) Mediterranean was for many years mainly a source of cheap labour, the interest of German society in the area was by and large limited to tourism and cultural aspects. Hence, domestic constituencies' demands on the executive to adopt active policies towards the southern Mediterranean were negligible.

Hand in hand with the changes in Germany's internal and external environment and the erosion of the rationale for its external foreign policy at the beginning of the

1990s came a gradual reformulation of the country's *Selbstverständnis* in foreign policy. This was initiated in particular by the conservative–liberal government under the leadership of Helmut Kohl and the subsequent red–green coalition led by Gerhard Schroeder. Interestingly, in the framework of the so-called out-of-area debate that took place in the early 1990s there was a widespread consensus among the country's political left that it should oppose the governing parties' support for the idea of engaging Germany in peace-keeping and peace-building operations outside Germany, with a view to its assuming greater and, most of all, global responsibilities. Gradually, heavily influenced by their party leaders, who had finally understood the growing international demands, the Social Democrats changed their position in 1992. The constitutional court's ruling of 1994, considering out-of-area missions legitimate provided that they were preceded by a parliamentary decision, prepared the ground for the U-turn by the political left after it assumed power in 1998 and after Chancellor Schroeder risked a vote of confidence in November 2001.[2]

This shift from externally imposed foreign policy navel-gazing to global activism was the result of both external and domestic demands and was intertwined with the gradual build-up of the EU's common foreign and security policy, but it was also undeniably used by Kohl, Schroeder and the former foreign minister Fischer in their aspirations to sharpen and increase their own political profile beyond the realm of domestic politics (Clough, 1998; Schröder, 2006). Undoubtedly, the developments mentioned above, the EU's decision to make the Mediterranean an area of strategic importance, the numerous terrorist attacks in various southern Mediterranean cities over recent years (some of which claimed the lives of German citizens), the civil war in Algeria, the failure of the Madrid Peace Process, and the learning process among parts of the German foreign policy establishment as a result of its membership of the Euro-Mediterranean Partnership (EMP) contributed to the fact that the Mediterranean became an issue of post-unification German foreign policy (Perthes, 1998: 1). Nonetheless, as exemplified by the programmes of the German EU presidencies in 1999 and 2007, which hardly mentioned the Mediterranean at all, the relevance of this context did not generate a single German 'Mediterranean policy' as such. Moreover, in spite of Germany's participation in the Schengen agreement, which supposedly triggers greater sensitivity to developments in the south, in Germany, both on the societal and political level, the 'neighbourhood' is still mainly associated with central and eastern Europe. The Mediterranean as a foreign policy arena is also subordinate to Germany's political, economic and socio-cultural neighbourhood, which extends to the US, owing to more than 60 years' close bilateral co-operation.

Nevertheless, in spite of these observations, the German government played a crucial role in upgrading Euro-Mediterranean relations in the run-up to the Barcelona Conference in 1995. This engagement was, however, rather the result of an intensive bargaining process, involving the practices of issue-linkage and favour-exchange, mainly between Spanish prime minister Felipe González and Chancellor Kohl, at the end of which Germany accepted the initiation of the EMP in exchange for Spain's support for German-inspired plans to start the EU accession process for the central and eastern European reform states (Gomez, 2003). Kohl's principal position at the time was to give free trade priority over political co-operation and prevent both the closer association of southern Mediterranean partners with the EU and greater financial assistance. This has represented a consistent stance of all German governments ever since. Using the instrument of coalition-building, in particular with governments of other non-Mediterranean member states, Germany has only been partly successful in ensuring this line and has even occasionally displayed a contradictory attitude in the wake of free trade negotiations by acting in a highly protectionist fashion (Schumacher, 2005: 229). While Kohl and his British counterpart Major were instrumental in ensuring acceptance of the Delors II package in 1992, which in turn prevented even greater financial assistance under the MEDA I programme three years later, and Chancellor Angela Merkel successfully orchestrated an anti-Mediterranean Union (UM) coalition in the early months of 2008, German governments failed to prevent either the incorporation of the southern Mediterranean into the European Neighbourhood Policy (ENP) and increased financial assistance within the framework of the latter or the granting of advanced status to Morocco in the autumn of 2008.

In addition to Germany's long-standing commitments in the area of development assistance to the southern Mediterranean, two policy areas stand out – trade and Germany's special relationship with Israel and, thus, its growing interest in contributing to conflict resolution in the Middle East.

Soon after the Second World War, the governing elite, with the support of the western bloc, was already focusing on the creation of a market economy and the pursuit of a liberal export-oriented trade policy to generate an international network of interdependence, which was also to be used in the context of (West) German governmental efforts to overcome the partition of Germany. This rationale and the end of the colonial period led to the gradual establishment of diplomatic relations and an intensification of bilateral trade relations with all southern Mediterranean countries. As part of this development, German industry, represented by the Federation of German Industries (BDI), along with the German–Arab Association and subsequent German–Arab economic forums, became instrumental over the years in intensifying these trade links and, through their policy demands, contributed to the fact that nowadays Germany is among the most important trading partners of all southern Mediterranean countries (Miller and Mishrif, 2005). In conjunction with the fact that existing Euro-Mediterranean association agreements predominantly contain trade stipulations, the above group's consistent lobbying impacted seriously on the actions of all German governments in the field of the promotion of political reform. In effect, such issues as the strengthening of human rights, good governance and democratization in the South – in purely practical political terms – became subordinate to well-defined trade interests (Schumacher, 2009: 182).

The role of personalities in foreign policy making was particularly obvious in Germany's recent pro-active engagement in the resolution of the Israeli–Palestinian conflict, as former foreign minister Fischer – driven by personal ambition, the moral imperative of Germany's past, and demands by many Arab governments to contribute to a just and peaceful resolution to the conflict – made the issue a priority in German foreign policy. His seven-point 'Idea Paper' of April 2002 and his second four-page Middle East peace initiative of late 2002, though unsuccessful, became the basis of the EU's position and to some extent the blueprint for the roadmap for peace. The latter was adopted by the Middle East Quartet in December 2002. In the person of foreign minister Steinmeier, though somewhat less prominently, the Foreign Ministry has continued along Fischer's path and played a key role in pushing Germany's first ever military operation in the Middle East – in the context of the United Nations Interim force in Lebanon (UNIFIL II) Maritime Task Force – through Parliament.[3] Since Guido Westerwelle replaced Steinmeier in late 2009, the Foreign Ministry has however become even less active with respect to the Middle East and North Africa, although in May 2010 it initiated the German–Palestinian Steering Committee, which aims to facilitate institution-building in the Palestinian Territories. In contrast to Fischer and Steinmeier, Chancellor Merkel has picked up on demands from various Jewish communities in Germany and the Israeli government of prime minister Olmert and displayed a more Israel-friendly position – thereby jeopardizing the presently more balanced perception in the southern Mediterranean and Arab world of Germany as an impartial negotiator in the conflict.

*Defending the Unwanted: Germany vs. France*

As far as the UM/UfM is concerned, Merkel (and, thus, the German government) acted as a veto-player, at least from autumn 2007 until the Brussels European Council of 13/14 March 2008. German opposition to a project that was originally destined to adopt the form of a non-EU co-operation framework, excluding the majority of EU member states, was not rooted in any societal demands or pressures. Nor was the German government informally tasked by other EU member states' governments to take the lead in opposing French President Sarkozy. Interestingly, German industry also kept a rather low profile during the months preceding the Brussels summit of 13/14 March 2008, in spite of the fact that German business would have lost a potential opportunity to expand its market share in the South if the original French plans had been successful. In fact, the BDI only published a position paper after Merkel had finally managed to convince Sarkozy, at their bilateral meeting in Hanover in early March 2008, to abandon his exclusionary plans and it had been guaranteed that the new project would include all EU member states and even all non-EU Mediterranean riparians. In this document the BDI generally welcomed the new initiative, considering the Mediterranean 'an interesting market thanks to a growing dynamism and much untapped potential' but, even so, pointed to ten major challenges that the UfM would need to address in order to become an economic success.[4]

In the run-up to the Brussels summit, none of these concerns were ever raised by Chancellor Merkel or foreign minister Steinmeier, who was considerably less outspoken in his criticism of the UM notion. Instead, the point of departure for the Chancellery's criticism of the Sarkozy initiative, which grew considerably throughout 2007 and reached a climax in December 2007, was a carefully chosen argument intended to shift the focus to the EU level and thus away from what were, in fact, purely power-oriented considerations: after months of deliberate restraint and silence, providing the Elysée with ample space to abandon the idea of a UM at a very early stage, Merkel argued that the creation of a UM that included only Mediterranean riparians had the potential to set in motion gravitational forces within the EU that in turn could generate a process of fragmentation and, eventually, disintegration. She reminded Sarkozy, and hence all other EU governments, that the use of EU funding for the pursuit of exclusively national interests could not be justified (Schmid, 2008). Fully aware that these arguments would raise concern among the governments of other EU member states, she hardly missed an occasion to make her message heard, with the aim of bringing potentially diverse perceptions in line with one another and thereby signalling to other potential veto-players that Germany was determined to oppose any proposal based on exclusion of some EU member states. Obviously, this strategy was intended to portray Merkel as acting in defence of the 'common good', i.e. the very existence of European integration and EU-European commonality. On the other hand, the rationale underlying this strategy was to prevent France from becoming *primus inter pares* in European foreign policy matters and thereby undermining Germany's role as the leading actor within the EU, and to preclude a resurgence of French colonial ambitions. Another layer was added to this multi-level game by the incorporation of the growing concern among the German foreign policy elite that French President Sarkozy's ignorance of long-standing bilateral communication and co-ordination channels had the potential of seriously affecting the Franco-German alliance – after all, a cornerstone of post-war German foreign policy and, due to the deep degree of mutual interdependence, almost a 'domestic' issue.

Before the French–German meeting in Hanover, the 'Appel de Rome',[5] adopted by the prime ministers of Italy and Spain and the French president on 20 December 2007, gave a good indication of the first impact that Merkel's warnings had had in other EU capitals. It also showed that German policy entrepreneurship was destined to stop Sarkozy and, hence, the informal German-led coalition-building that had already started in the background had finally begun to bear fruit. By downgrading the proposal from a Mediterranean Union to merely a Union *for the* Mediterranean and by suggesting that *all* EU member states should attend the Paris summit of July 2008, the dynamics had changed and the Chancellery was using this momentum to play its cards one after the other, thereby gradually increasing the pressure on the Elysée. It was almost a logical step for Merkel to go beyond her repeatedly raised concerns and open yet another front that would make it impossible for Sarkozy to push through his exclusive plans. Encouraged by Merkel, this front was opened in the form of a policy speech by the newly elected Polish prime minister Donald Tusk on 23 November 2007, in which he mentioned that Poland should 'participate in shaping the eastern dimension of the EU through the development of relations with Ukraine and Russia'.[6] From the Chancellery's perspective, the beginning of an intra-EU discourse, emerging simultaneously, on the possible need to establish an 'Eastern European Union' and the linkage of two possibly emerging policy frameworks for Europe's most sensitive neighbourhoods finally ensured the attention of all EU governments. Moreover, it opened new avenues for Merkel to achieve what she had already announced in her speech before the European Parliament on 17 January 2007, namely that greater attention should be paid to eastern Europe (Schumacher, 2008).

From a conceptual point of view, it became evident that once Germany had given up the 'wait-and-see' approach that prevailed throughout most of 2007, it can be said to have acted as a reverse policy entrepreneur. Merkel was determined to defy Sarkozy's uncoordinated, unilateral plans and frame the issue of Euro-Mediterranean co-operation in a fashion that would make other EU member states concur with it and generate, ideally, a Europe-wide debate. In this context, Germany's foreign policy behaviour displayed features of both a norm entrepreneur and a believer entrepreneur. It countered France's plans for a UM/UfM with the initiation of a process of norm emergence, which was based on the acknowledgement that Euro-Mediterranean relations may need to be reinvigorated and the notion that any such revitalization must take place within a truly Euro-Mediterranean setting. This allowed the German government to position itself at the centre of the debate, define what it considered appropriate, identify itself fully with the 'counter norm' it aimed to promote, i.e. full Europeanization of any potential initiative and, finally, work on persuading other EU governments to embrace this norm. Interestingly, from the moment it was certain that Germany's counter norm would be assimilated and become the common norm that would underpin future Euro-Mediterranean relations, the German government re-assumed its original role as a loyal, albeit passive player on the issue of Euro-Mediterranean co-operation. It adopted a more assertive position only in more technical discussions, such as those in the second half of 2008 relating to the future structure of the UfM secretariat and the mandate of its deputy secretary-general, which, in the view of the German government, should involve horizontal responsibilities. It maintained this attitude throughout 2009 and, in an informal communication in early 2010, called upon other EU governments to adopt a minimalist approach to the UfM secretariat's budget. Apart from these occasions, however, it has taken a back seat ever since, as it has watched the UfM, and thus Euro-Mediterranean relations, which it never favoured anyway, plunge from one crisis to another.

## Poland

*Polish Foreign Policy in the Post-Cold War Period: Mediterraneanization through Europeanization?*

Undoubtedly, the defining moment for current Polish foreign policy was the fall of the Berlin Wall and the subsequent collapse of the Warsaw Pact. As the country was forced by the latter to surrender its autonomy in the field of foreign policy to the Soviet Union, the termination of the Pact in July 1991 led to a situation in which Polish society and the political elite – old and new – had to embark on a discourse over the future course of the country's foreign policy. With the election of Lech Walesa in December 1990 as president of the Third Polish Republic, after decades of totalitarianism, it soon became obvious that this discourse was less about whether

Poland should or should not develop a western orientation than about the extent to which this general orientation was synonymous with full or just partial integration into Euro-Atlantic structures (Prizell et al., 1995). Eventually, societal and political consensus emerged in favour of full integration into both NATO and the EU. Although this process suffered repeated setbacks, Poland joined NATO in 1999 and the EU in 2004.

Participation in Euro-Atlantic structures and the process of European integration set the internationalization of Polish post-Cold War foreign policy in motion, embedding the country in new co-operation structures and thus increasing the degree of interdependence between it and its partners in this newly evolving co-operation. This development occurred, however, at the expense of the relations with (Mediterranean) countries that Poland had developed in the Cold War, not least for ideological reasons. While bilateral relations were established and maintained in particular with Syria, Algeria and Libya, the relative importance of these relationships declined as a consequence of the diversification of Polish foreign policy and the growing concern among both governmental and societal actors about Poland's mainly non-democratic eastern neighbourhood. Unsurprisingly, this development passed almost unnoticed and was never the subject of domestic debate among the political elite, the media or other constituencies and lobby groups, e.g. Polish industry, the Poland Import Export Chamber of Commerce, the Polish Information and Foreign Investment Agency or the Polish agricultural lobby. The absence of specific policy demands from domestic actors with respect to the southern Mediterranean region, noteworthy since the creation of the Third Republic, is thus even more blatant in Poland than in Germany (Wojna, 2008).

This situation, i.e. the absence of domestic Mediterranean-related policy supplies and thus domestic preference formation, remains almost unchanged in spite of Poland's EU membership and its corresponding participation in the EU's Euro-Mediterranean co-operation framework. At government level, however, the Europeanization of Polish foreign policy is discernible to the extent that every single Polish government, in the context of EU membership, has officially committed itself to the EMP, now the UfM, and supports the creation of the Euro-Mediterranean Free Trade Area. Certainly, as in the case of Germany, exposure to Euro-Med practices and participation in sectoral co-operation programmes has led to greater, albeit still underdeveloped, sensitivity among Polish decision makers as well as increased awareness of the socio-economic and political developments in the southern Mediterranean. In recent years, as a result of this socialization process, an increasingly firm grasp of the market potential of southern Mediterranean countries and an awareness of the need to diversify energy supplies, Polish governments have gradually started to reinvigorate their relationships with some of the country's former ideological allies in the Mediterranean. The leading government actor in this regard is the Ministry of the Economy: as a result of both a visit by a Polish government delegation to Algeria in 2006 and a bilateral meeting between the economy minister Piotr Grzegorz and the Algerian energy and mines minister in January 2007 in Warsaw, it initiated a memorandum on co-operation. This was intended to lead to the strengthening of bilateral economic relations, particularly in the fields of energy, mining, telecommunications, transport and construction.[7] In the light of Algeria's position as the third most important market for Polish exports in Africa and its vast energy resources, the re-intensification of relations is a natural development.

Such an explanation, however, does not apply to Syria. Yet, on 5 March 2009, for the first time in 20 years, under the leadership of deputy economy minister of the economy Adam Szejnfeld, together with the Polish Chamber of Commerce, the Ministry of the Economy held a Polish–Syrian business forum in Warsaw to identify areas of future co-operation. It was preceded by bilateral negotiations and the conclusion of an agreement to set up a Poland–Syria Business Board. As the meetings mainly revolved around issues such as the operation of special economic zones in Poland, co-operation in the field of food processing, construction, infrastructure and utilities, their underlying rationale is simply related to the government's objective to explore new markets at a time when the European single market is in recession, and thus increase the bilateral trade balance, currently amounting to approximately $82m.[8] However, sensitive issues pertaining to Syria's role in the Israeli–Palestinian conflict, its relations with Hamas and Hizballah, and its special relationship with Iran – all of which are of utmost concern to the EU – were never addressed by the Polish government during the meetings. Indeed, since 1989, Polish foreign policy has officially been committed to the protection of fundamental rights, the rule of law and democracy but, apart from being the cornerstones of Poland's policy in international frameworks, these principles have been addressed in the context of Poland's relations with its eastern neighbours and, most recently, the Russian–Georgian War in 2008, but have never been the subject of any direct intergovernmental encounter with any of the EMP's southern partners.

The formation of government actors' interest in the southern Mediterranean is only discernible to the extent that the 'Strategy for Poland's Development Co-operation' (Polish Ministry of Foreign Affairs, 2003), adopted by the Polish government in October 2003, singles out the Palestinian Territories as recipients of Polish Official Development Assistance (ODA), to be transferred either directly via the Polish Representation Office (opened in 2004 in Ramallah) or UNRWA (United Nations Relief and Works Agency for Palestine Refugees in the New East). Polish aid to the Palestinian Authority (PA) increased from €130,000 in 2005 to €500,000 in 2007, owing mainly to a decision taken by the EU's General Affairs and External Relations Council in April 2006 to meet the basic needs of the Palestinian population and address the deteriorating humanitarian situation (Kolarska-Bobinska and Mughrabi, 2008). It is questionable, however, whether Poland's development assistance to the PA can be considered a sign of Polish ambitions to assume a political role in the region. As is argued elsewhere, the position prevailing among government officials seems to be that 'development aid grants visibility' and thus is not a direct result of an intrinsic policy but rather a vehicle through which other political objectives not related to the Palestinian Territories and/or the Israeli–Palestinian conflict can be achieved (Kolarska-Bobinska and Mughrabi, 2008: 15). Undoubtedly, it is in this light that the 'Polish Strategy towards Non-European Developing Countries', containing one chapter on North Africa and the Middle East, has to be read and, secondly, it is against this background that Poland has strengthened its military presence as part of UNIFIL II. Polish engagement in the southern Mediterranean, be it in the context of development assistance or in peacekeeping missions, does not stem from an explicit 'Mediterranean agenda' but is rather the result of Polish governmental and societal desires to secure the country's political and economic interests, and the new responsibilities imposed upon it by EU membership (Kolarska-Bobinska and Mughrabi, 2008: 14).

*Poland and the Union for the Mediterranean: From Bystander to Favour-Exchanger*

When, in late November 2007, prime minister Tusk declared that Poland should act as a norm entrepreneur and adopt an even more pro-active stance within the EU to facilitate, in particular, the latter's relations with Russia and the Ukraine,[9] this announcement was in line with Poland's post-Communist foreign policy objectives, its long-standing considerations regarding regional stability and interdependence, various demands from domestic economic actors and, given Poland's recent history and geographical location, the country's broad societal attitudes. Having been in office for just seven days at the time of the speech, Donald Tusk refrained from making any reference to the UM/UfM, as positive and negative comments alike would have generated criticism either at home or in France – the latter being one of Poland's key strategic partners in the EU and a member of the Weimar Triangle. Instead, the newly formed Polish government, already aware of the differences gradually surfacing within the EU over the future course of the EMP, very quickly identified the intersection of interests and attitudes between the Polish and German (and other non-Mediterranean EU member states') domestic constituencies and, without stressing the fact explicitly, sided with the German Chancellery in its opposition to the creation of the UM. Although the creation of the UM/UfM was never the subject of public debate in Poland, or even discussed in Parliament, Tusk picked up on the general sentiment that such a union would possibly require greater financial and political involvement by all EU member states, which in turn was perceived as a development that could have negative repercussions on the further development of EU policy towards eastern Europe and thus on Poland and Germany's ambitions in eastern Europe.

The existence of overlapping concerns and interests between Germany and Poland did eventually allow governments of other non-Mediterranean EU member states to formulate, albeit indirectly, their unease with the French initiative and gradually position themselves, ahead of the Brussels European Council of March 2008. The Swedish government, particularly in the person of foreign minister Carl Bildt, subscribed to Tusk and Merkel's principal argument that the stabilization and democratization of eastern Europe must not be forgotten in the debate over the UM/UfM and, though avoiding any official remarks in that regard, Bildt resorted to issue-linkage by linking Poland and Sweden's approval of an inclusive and cost-neutral UM/UfM to the creation of an Eastern Partnership.[10] From the perspective of the newly elected Polish government, the declared intention to propose an Eastern

Partnership at the Brussels European Council in May 2008 generated a multi-faceted win-set in that it would guarantee broad domestic support for a major policy initiative, guarantee that Poland's most pressing foreign policy concern would be elevated to EU level, guarantee that the UM/UfM could not arouse unwelcome distributional consequences in financial terms and, hence, ensure that the newly elected government would simultaneously achieve a number of objectives without directly offending any of its EU partners.

For a long time, Poland displayed the features of a veto-player that genuinely disliked the perspective of reinforced Euro-Mediterranean relations and greater emphasis on the southern neighbourhood in the EU's foreign policy agenda. In contrast to Germany, however, by keeping a low profile and not publicly opposing the French initiative, it passed the buck and never openly acted like a veto-player. Instead, it exploited Merkel's explicit and outspoken stance against the UM to create an action corridor allowing itself to act as a believer entrepreneur advocating its own foreign policy objectives. The Polish government showed a relatively low degree of flexibility on the foundations and shape of its policy initiative and appeared to be prepared to oppose Sarkozy at the 2008 Brussels summit, knowing, of course, that it would have the backing of at least Germany and Sweden. Irrespective of whether the government was really determined to take such a step, it is worth pointing out that the mere existence of the French proposal, in conjunction with Germany's opposition to it, generated the preconditions for potential side payments and that prime minister Tusk and his government instantly recognized/understood this emerging window of opportunity. Tusk's decision to refrain publicly from joining the choir of critical voices, leaving it to Germany, not known for its Mediterranean track record, to defend the EMP and acting rather like a favour-exchanger, was logical – and a natural consequence of the way the entire dynamic evolved. The fact that Poland has displayed some degree of engagement in the senior officials' meetings, in particular since their resumption in mid-2009, must not, however, be interpreted as a sign of sudden interest or concern for Euro-Mediterranean issues. Instead, it is simply an expression of a general Polish paradigm to make its voice heard in multilateral settings with a view to obtaining greater visibility.

**Hungary and the Czech Republic**

*Hungarian and Czech Foreign Policy after the End of the Cold War: A Little Mediterranean at Last?*

As in the case of Poland, the end of the Cold War marked the dawn of a new era for Hungary and the Czech Republic[11] and led to a total relaunch of their foreign policy. For decades, their location behind the 'iron curtain' determined/dictated/governed the ideological framework for their actions and interactions on the international level, and membership of the Warsaw Pact determined and, indeed, limited the scope of their foreign policy (Kun, 1993). The dissolution of the Soviet Union in 1991 and the disappearance of its omnipresent hegemon provoked a wide-ranging ideological overhaul of the foundations of their foreign policy and affected both countries' international outlook, their bilateral relations, and the degree and depth of their involvement in multilateral structures. In addition, of course, it opened new avenues for co-operation with countries from hitherto neglected world regions.

This applied in particular to relations with Israel. Both Hungary and the then Czechoslovakia were among the first countries to recognize the state of Israel, with Czechoslovakia recognizing Israel in 1948 and Hungary shortly afterwards.[12] The presence of large Jewish communities in both Hungary and Czechoslovakia, many of which emigrated to Israel, collective identities, a joint cultural heritage and, of course, the shoa were among the defining factors of the bilateral relationships. Due to this mix of variables, bilateral ties very quickly strengthened and, as far as Czech–Israeli relations were concerned, were even preceded by the supply of crucial military aid to Israel by the Czechoslovak government in mid-1948, in spite of a United Nations arms embargo. As a consequence of the worsening 'bloc confrontation' and the Soviet Union's growing political and military engagement in Egypt in the mid-1950s, both Czech–Israeli and Hungarian–Israeli relations started to deteriorate and, under Soviet pressure, eventually ceased in the wake of the Six Day War in 1967.

On 18 September 1989, however, Hungary re-established full bilateral relations with Israel and Czechoslovakia followed suit in early 1990. The consideration that the re-establishment of relations with Israel would be regarded favourably by the US and important European countries and help Hungary's and Czechoslovakia's bids for integration into transatlantic and European structures was certainly not a negligible motive for the two countries' governments. Yet, more importantly, the decisions of the Hungarian and Czechoslovak governments to re-establish relations as swiftly as possible were mainly influenced by the set of factors that had already guided Hungary and Czechoslovakia's almost instant recognition of Israel in the late 1940s. To date, relations have been close and governmental representatives of both the Czech Republic and Israel have repeatedly stated that the bilateral relationship is special.[13] In the latter case, this is reflected in the existence of 11 co-operation agreements, covering bilateral co-operation in areas as diverse as private sector industrial research and development, social security, health and medical sciences. In comparison, there are just five co-operation agreements between Hungary and Israel.

Although in the Czech Republic today there are only about 4,000 Jews, compared to around 100,000–130,000 in Hungary,[14] the Federation of Czech Jewish Communities has managed, over the years, to act as a small but influential domestic constituency. The Czech Republic's close links with Israel came most visibly to the fore in the Gaza War, in early 2009, when the then Czech EU Presidency published a statement on behalf of the EU declaring that Israel's 'Operation Cast Lead' was 'more defensive than offensive'.[15] While the Israeli media welcomed such a declaration and Israeli political circles interpreted it as *carte blanche* for further military action in the Gaza Strip, other EU leaders condemned the escalation of violence and called for an immediate ceasefire.[16] Interestingly, the Hungarian government was also among those calling for the earliest ending of the violence, although, like the Czech Republic, it pointed to the continuous provocations in the form of rocket attacks from the Gaza Strip and made clear that it supported Israel's right to self-defence.[17] Only the strong opposition that rapidly arose in other EU capitals to a statement that was overwhelmingly considered biased and out of touch with EU member states' positions led foreign minister Schwarzenberg to backtrack and adopt a more balanced standpoint. The Czech declaration, adopted just one day after the initial remarks of Foreign Ministry spokesperson Jiri Potuznik, pointed out that 'the indisputable right of the state to defend itself does not allow actions which largely affect civilians'.[18] This did eventually calm the waters but could not disguise the general attitude of the Czech intelligentsia or foreign policy elite. Hence, it was no surprise that at the end of April 2009 caretaker prime minister Mirek Topolánek, who was the first prime minister to visit the new right-wing Israeli government, once more defended Israel's attack on the Gaza Strip and even entered into a public dispute with European commissioner Ferrero-Waldner.[19] Ultimately, as a consequence of Belgian, Swedish and Portuguese opposition, the Czech EU Presidency was forced to accept that the upgrade in relations would have to be postponed and implemented only once the new Israeli government had explicitly signed up to a two-state solution, abandoned settlement activities and lifted the blockade of the Gaza Strip.[20]

Following the political changes of 1989–90 and the renaissance of Jewish life in Hungary, Zionist organizations are more visible and active in the political, public and cultural sphere than their counterparts in the Czech Republic. In contrast to the Czech Republic, however, their policy supplies and policy demands relate mainly to domestic issues and, lately, more to individual and collective efforts destined to reclaim expropriated property than to foreign policy matters as such. In addition to existing Jewish civil society organizations, the domestic constituencies in both Hungary and the Czech Republic that display at least some degree of interest in greater (commercial) ties with Israel and other southern Mediterranean countries are the various chambers of commerce, the two countries' investment and trade development agencies, and certain business associations. These associations mainly represent local industries in the areas of chemicals, machinery, mechanical devices, pharmaceuticals, the automotive sector (in the case of the Czech Republic) and the Hungarian oil and gas company MOL (in the case of Hungary). In fact, the latter two are the most powerful and influential lobbies in support of full exploitation of existing Euro-Mediterranean association agreements, possessing a considerable voice opportunity.[21]

In conjunction with the two countries' membership of the EU, and thus full adoption of its trade regime, economic ties with Israel have grown considerably in recent years. In the case of Hungary the trade volume in 2008 amounted to $463.3m, up from $311.9m in 2007 and, as far as the Czech Republic's trade exchanges with Israel are concerned, total trade in 2008 amounted to $327.2m, compared to $269m in 2007.[22] According to Israeli sources, Hungary has even become central and eastern Europe's biggest importer of capital from Israel, with annual investment capital of $200–300m flowing, above all, into the Hungarian real estate, industrial, infrastructure, and research and development sectors.[23]

During the Cold War, Hungary's and the Czech Republic's relations with Arab southern Mediterranean countries, in particular with states that were considered friendly, e.g. Algeria, Libya, Egypt and Syria, were based on ideological commonalities and proved to be highly important with respect to Hungarian and Czechoslovakian export commodities. As pointed out by Rozsa (2005: 5), however, 'these relationships were drastically scaled down if not abandoned with the transition, and as political attention turned elsewhere their basis became essentially economic'. Especially since both countries joined the EU and became members of the EMP, though also in view of the two countries' official policy doctrine that – in spite of close relations with Israel – a balanced approach towards the Israeli–Palestinian conflict needs to prevail, diplomatic relations with the Arab southern Mediterranean have steadily intensified again. As far as Hungary is concerned, they culminated in a recent announcement by the Hungarian foreign minister Balázs. On the occasion of a meeting with Arab ambassadors accredited to Budapest, he declared that the Arab world as a whole had a 'prominent place in Hungary's foreign policy' and that Hungary aimed 'to facilitate the further improvement of broad co-operation'.[24] Similar statements were made by Czech senior officials during the Czech EU Presidency.[25] Yet, in spite of the opening of diplomatic representations in all southern Mediterranean countries, in both cases the action does not match the rhetoric, as neither Hungary nor the Czech Republic have ever been able to play a significant role in the Middle East Peace Process (MEEP) or other conflicts such as the Western Sahara conflict or the July War of 2006. The original statement by the Czech EU Presidency in early January 2009 on the outbreak of violence in the Gaza Strip, although withdrawn the next day, is still being referred to by the Arab media as evidence of the Czech Republic's bias in favour of Israel (Zaidi and Sami, 2009).

Exposure to and participation in EU and transatlantic co-operation initiatives with the Arab southern Mediterranean have influenced the Czech Republic's security strategy, which emphasizes the 'great importance' of relations with the countries of the Middle East and North Africa, not least due to the putatively 'high-risk' character of the region (Ministry of Foreign Affairs of the Czech Republic, 2004). In Hungary's case, integration into Western structures has led to the adoption of a policy paper, already elaborated by the Foreign Ministry in 2000, suggesting a concept for a possible southern Mediterranean policy.[26] In practical terms, however, these documents did not make much of a difference as both Hungarian and Czech policies – with the notable exception of some of the statements made during the Czech Republic's EU Presidency – adhered more or less strictly to the official EU policy line. They also failed to contribute to a public discourse about whether and to what extent the southern Mediterranean should play a greater role in their foreign policy agendas. Apart from Operation Iraqi Freedom, in which both countries were involved, and the media coverage of the July War 2006 and the Gaza War 2008/09, the Mashreq and the Maghreb are seldom the subject of public or parliamentary debate. In this vein, given the absence of sizeable Muslim communities in both countries, it is not surprising that Hungary's and the Czech Republic's participation in the EU Police Co-ordinating Office for Palestinian Police Support (EUPOL COPPS) and Hungary's contribution to the EU Border Assistance Mission at the Rafah Crossing Point (EUBAM Rafah) were not subject to parliamentary scrutiny or in-depth media coverage.

Due to the low degree of importance attributed to the Arab southern Mediterranean, in addition to the two countries' post-Communist political culture, in which foreign policy matters are hardly an issue of domestic interest and debate, the growing commercial interdependence with the Arab southern neighbourhood remains widely unnoticed. In particular, Hungary's trade ties with the entire region are expanding, as witnessed by the fact that in recent years Hungarian exports to the Middle East and North Africa have grown twice as fast as the country's average total exports.[27] In contrast, the Czech Republic's trade with the Arab southern Mediterranean is relatively consistent, without major fluctuations. In 2008 the trade volume amounted to CZK 13,455.2m (Ministry of Industry and Trade/Czech Statistical Trade Office, 2009). The exceptions are Egypt, the only country in the south where the Czech National Trade Promotion Agency of the Ministry of Industry and Trade is represented, and Morocco, with which total trade is growing steadily. Apart from the latter, which is one of the Czech Republic's main suppliers of agricultural products, regular trade surpluses exist with all other Arab southern Mediterranean countries and in 2008 four of them even ranked among the Czech Republic's 65 most important trade partners worldwide.[28]

*Hungary, the Czech Republic and the Union for the Mediterranean: Potential Veto-Players or just Unhappy Laggards?*

The Hungarian and Czech government elite also displayed a rather negative attitude towards France's UM initiative and regarded the emerging dynamics with great concern. In contrast to Germany and Poland, however, this opposition did not result in any (counter-)policy entrepreneurship and did not receive any noteworthy media coverage. Hungary and the Czech Republic did not deliberately try to exploit the lack of agreement over the future course of Euro-Mediterranean relations. They did not even adopt the role of 'honest brokers', nor did they make any visible effort to undermine Sarkozy's plans.

In both countries, the overall mood in the prime minister's office and the foreign ministry oscillated between scepticism and outright opposition. The followers of either mindset hinted at the re-emergence of the odd intra-EU allocation debate of the early 1990s over the extent to which financial assistance should be given to Europe's eastern and southern neighbourhood, the negative implications the UM may have on European neighbourhood policy or the EU's interest in stabilizing eastern Europe and, finally, the potential exclusiveness of the UM. Although these negative attitudes were for intra-ministerial consumption and did not have an effect on any potential entrepreneurial policy behaviour or public opinion, they underpinned the overall attitude of both governments and contributed to the fact that Hungary and the Czech Republic alike can be described as unhappy laggards in the overall process leading to the UfM. Undoubtedly, in view of the two countries' foreign policy priorities their generally negative attitudes were genuine, but in the course of time they were complemented by certain strategic considerations.

The prevailing policy doctrine in both countries, that EU membership implies solidarity, can partly explain the inaction by their governments. Yet the refusal to join the ranks of Merkel and Tusk and make their dissatisfaction public was also due to an inherent sense of being small EU member states with a low voice opportunity, in addition to the nature of their bilateral relations with France. During prime minister Topolánek's visit to Paris in October 2007, he and Sarkozy established the Franco-Czech Partnership, which was intended to intensify political, economic, cultural, linguistic, university, scientific and technical co-operation. Similarly, the Hungarian prime minister Ferenc Gyurcsány, albeit only in May 2008, signed a document initiating a 'strategic partnership' between Hungary and France (Rozsa, 2009). As far as Hungary is concerned, the establishment of a bilateral strategic partnership can be considered a reward and a side payment in exchange for Gyurcsány's moderate public stance on the French initiative. Having managed to silence the many critical voices within his administration, the Hungarian prime minister succeeded in balancing Franco-Hungarian and Franco-German relations, both of which are of major importance to Hungary. Of course, this included not antagonizing either Merkel or Sarkozy.[29] Surely, this balancing act, widely ignored by potentially relevant domestic constituencies, was a deliberate and strategic choice aimed at averting a situation in which Hungary would find itself squeezed between French unilateralist bulldozing and German opposition to it.

A similar approach is discernible in the Czech reactions. Prime minister Topolánek was wholehearted in his scepticism towards a project that threatened to divert financial and political resources away from eastern Europe and was likely to jeopardize Turkey's EU membership prospects (an issue widely supported by the Czech Republic) but, as in the case of his Hungarian counterpart, he never expressed it publicly. Although the opposition and domestic constituencies did not show any interest in the intra-EU discourse over future Euro-Mediterranean relations, fears that the adoption of an outspoken position either in favour or against the UM/UfM would be detrimental to the Czech Republic's relations with other EU member states, most notably Germany, France, Poland and the United Kingdom, determined the course of governmental action. Undoubtedly, these concerns were complemented by considerations that, if the Czech EU Presidency in the first half of 2009 was to be a success, any confrontation with important EU member states needed to be avoided. Hence, Topolánek's decision to follow the Hungarian example and let other EU governments take centre stage was a logical consequence of the external constraints the Czech Republic was and is exposed to.

After the Brussels European Council summit of March 2008, the Czech government did leave itself open to ridicule when its Foreign Ministry leaked a transcript of a meeting between Topolánek and Sarkozy that took place at the end of October 2008, just days after the then ruling Civic Democrats of prime minister Topolánek lost their majority in the Czech Senate. According to this document, the Czech government's decision to allow France to keep the co-Presidency of the UfM even during the Czech EU Presidency was simply part of a package deal offered by Sarkozy. Under the threat of an election defeat and faced with early elections in 2009, the domestically weakened Czech prime minister did not hesitate to surrender the Czech co-Presidency to France. This was motivated, in particular, by the fact that Sarkozy promised to refrain from interfering in eastern European matters, to acknowledge Topolánek as *the* leader of central and eastern European EU member states, and to facilitate President Barack Obama's presence at the EU–US summit in Prague in April 2009. Such concessions could have given the deeply entangled Czech leader much-needed momentum if they had ever materialized.[30]

Interestingly, while Hungary has maintained its general attitude towards the UfM and still pursues an approach that could be best described as engaged non-engagement, the Czech caretaker government of prime minister Fischer has moved away from the general approach of his predecessor. In early 2010 he joined the ranks of Germany, the United Kingdom and Sweden and opposed any decision destined to lead to a fairly generous budget for the UfM Secretariat.

## Conclusions

This study has shown that although Germany, Poland, Hungary and the Czech Republic share essential foreign policy objectives and concerns with respect to both eastern Europe and the southern Mediterranean, as well as similar geopolitical preconditions, their response to Sarkozy's plans to establish the UM differed considerably. In spite of the existence of at least a few relevant domestic constituencies in each of the four countries that could have generated concrete policy supplies and put pressure on their governments to act in a way that would correspond with their policy demands, they did not have any influence on the governments' responses. Mainly due to their different bargaining powers within the EU, though also to simple power considerations, governments – in the case of Germany and Poland – assumed either the role of 'reverse' or 'counter-policy' entrepreneurs, thereby acting as veto-players or favour-exchangers, or, as in the case of Hungary and the Czech Republic, of unhappy laggards, linking genuine disagreement with Sarkozy's exclusionary plans with (minimal) efforts to generate at least some kind of side payment without antagonizing anybody.

Despite increasing awareness among the political and economic elites in all four countries of the importance of North Africa and the Mashreq, it can be seen that the southern Mediterranean and, hence, the UfM do not feature very highly on their foreign policy agendas. Apart from the fact that the UfM suffers from severe structural flaws, many of which are responsible for its poor performance, this situation and the fact that the governments of all four countries succeeded in turning the debate on the UM/UfM to their own advantage are the main reasons why Germany, Poland, Hungary and the Czech Republic have assumed a low profile since 1995 and 2004. This is, however, not without possible consequences for Euro-Mediterranean relations and EU foreign policy making in general, both of which are already suffering considerably from the existence and pursuit of heterogeneous foreign policy interests and non-negligible power calculations.

## Acknowledgements

The author would like to thank Colin Archer, Erzsébet Rozsa, Sharon Pardo and officials from the foreign ministries of Germany, Poland, Hungary and the Czech Republic. The sections on Germany and Poland draw on a previous study by the author, entitled Explaining Foreign Policy: Germany, Poland and the

United Kingdom in Times of French-inspired Euro-Mediterranean Initiatives, published in 2009 in *Hellenic Studies/Études Helléniques*, 17(2), pp. 205–238.

## Notes

[1] While West Germany was among the original signatories of the Schengen agreement, Poland, Hungary and the Czech Republic abolished border controls on 21 December 2007.
[2] See http://www.euractiv.com/en/security/schrder-wins-confidence-vote/article-114368
[3] Interestingly, after the Liberal Party had joined the CDU in forming a coalition in the wake of the parliamentary elections in September 2009, it voted in favour of extending the Bundeswehr's participation in UNIFIL II for another six months. During its time in opposition it opposed the extension of the Bundswehr's mandate.
[4] See http://www.bdi-online.de/Dokumente/Internationale-Maerkte/Mittelmeerunion_Position_engl.pdf
[5] See http://www.elysee.fr/documents/index.php?mode=view&lang=fr&cat_id=1&press_id=821
[6] See http://www.apcoworldwide.com/content/PDFs/112307-Tusk-speech.pdf
[7] See http://www.mg.gov.pl/English/News/Polish-Algerian+talks+on+economy.htm
[8] See http://www.mg.gov.pl/English/News/New+prospects+for+economic+cooperation+with+Syria.htm
[9] See http://www.kprm.gov.pl/english/s.php?id=1413
[10] For more on the Eastern Partnership, see Shapovalova (2009), and Jesien et al. (2008). See also http://cria-online.org/7_3.html
[11] The Czech Republic came into being after the dissolution of Czechoslovakia in 1993. The term 'Czech Republic' is used henceforth as far as the time after the dissolution of Czechoslovakia in 1993 is concerned. The term 'Czechoslovakia' applies with respect to the pre-1993 period.
[12] Czechoslovakia recognized Jewish nationality shortly after the establishment of the Czechoslovak state in 1918.
[13] See, for instance, Czech foreign minister Karel Schwarzenberg's remarks at the Israeli Council on Foreign Relations on 26 November 2007, published in *the Israel Journal of Foreign Affairs*, II(1) (2008), pp. 123–127. A speech in similar style was given by Hungarian foreign minister Kinga Göncz on the occasion of a reception to commemorate the 60th anniversary of the state of Israel. See http://www.mfa.gov.hu/kum/en/bal/actualities/visits_and_events/GK_IZR_eng_080513.htm
[14] Figures vary considerably due to intermarriage and emigration. It is estimated that there are an additional 10,000 to 15,000 unregistered Jews living in the Czech Republic. According to Rozsa (2005, p. 11), Hungary is home to the third largest Jewish community in Europe.
[15] See *Haaretz*, 4 January 2009, 'EU Presidency: Israel ground op in Gaza "defensive not offensive"'.
[16] See 'EU seeks Gaza role despite conflicting views', 5 January 2009, http://www.euractiv.com/en/foreign-affairs/eu-seeks-gaza-role-despite-conflicting-views/article-178272
[17] See the declaration issued by the Hungarian Ministry of Foreign Affairs on 29 December 2008, entitled 'Hungary considers important the earliest ending of Gaza strip military activities'.
[18] See 'EU seeks Gaza role despite conflicting views', 5 January 2009, http://www.euractiv.com/en/foreign-affairs/eu-seeks-gaza-role-despite-conflicting-views/article-178272
[19] See EUobserver, 27 April 2009, 'Czech EU Presidency splits with EU Commission over Israel': 'I consider the statement by Benita Ferrero-Waldner ... to be really hasty and at this given moment I would not really attribute to it more weight than just a statement by a commissioner. The action plan continuation is a political decision that is to be made by the European Council, and I am still the president of the European Council and I should know something about it.'
[20] See statement of the EU Council on the occasion of the 9th EU–Israel Association Council. Interestingly, in a joint press conference with Israeli foreign minister Lieberman summarizing the results of the Association Council, Czech foreign minister Kohout stressed the fact that no freeze in the upgrade of EU–Israel relations would occur.
[21] The term 'voice opportunity' was introduced by Joseph Grieco (1995).
[22] See State of Israel, Press Release 'Israel's Foreign Trade by Country – October 2009', 17 November 2009. Both countries generated trade surpluses, which in the case of Hungary amounted to $114.3m in 2008 and $79.3m in 2007, and in the case of the Czech Republic $4m (2008) and $3.8m (2007).

[23] See *Diplomacy & Trade*, 30 September 2009, 'Hungary, the region's biggest recipient of Israeli investments'.

[24] See Ministry of Foreign Affairs of the Republic of Hungary, 'The Arab world has a prominent place in Hungary's foreign policy – Péter Balázs met Arab ambassadors', 29 October 2009.

[25] Summary of the speech of the president of the EU General Affairs and External Relations Council Karel Schwarzenberg at http://www.eu2009.cz

[26] See Koncepció Magyarország dél-mediterrán politikájára [Concept for the Southern Mediterranean Policy of Hungary], August 2000, at http://www.kum.hu/Strategy/magyar/medit-m.htm and http://www.kulugyminiszterium.hu/kum/hu/bal/Kulpolitikank/Biztonsagpolitika/Nemzeti_biztonsagi_strategia

[27] See http://www.mfa.gov.hu/kum/en/bal/foreign_policy/hungary_in_the_world/

[28] Among the Czech Republic's global trade partners in 2008, Egypt ranked 52nd, Morocco 59th, Algeria 62nd, and Tunisia 65th (ibid.).

[29] During Sarkozy's visit to Hungary in 2007, his first stop in central and eastern Europe as French president, he praised prime minister Gyurcsány as the 'most European and most reliable statesmen' in central Europe (ibid.).

[30] Transcript of the meeting between French president Nicolas Sarkozy and Czech prime minister Mirek Topolánek, 31 October 2008.

# References

Clough, P. (1998) *Helmut Kohl. Ein Porträt der Macht* (Munich: dtv).

Gomez, R. (2003) *Negotiating the Euro-Mediterranean Partnership. Strategic Action in EU Foreign Policy?* (Aldershot: Ashgate).

Grieco, J. (1995) The Maastricht Treaty, economic and monetary union, and the neorealist research programme, *Review of International Studies*, 21(1), pp. 21–40.

Jesien, L. et al. (2008) Eastern Partnership – strengthened ENP co-operation with willing neighbours, *PISM Strategic Files*, Warsaw.

Kolarska-Bobinska, L. & Mughrabi, M. (2008) New EU member states' policy towards the Israeli–Palestinian conflict: the case of Poland, *EuroMeSCo Paper* 69, Warsaw.

Kun, J. (1993) *Hungarian Foreign Policy: The Experience of a New Democracy* (Santa Barbara, CA: Praeger Publishers).

Miller, R. & Mishrif, A. (2005) The Barcelona Process and Euro-Arab economic relations 1995–2005, *The Middle East Review of International Affairs*, 9(2), pp. 94–108.

Ministry of Foreign Affairs of the Czech Republic (2004) *Security Strategy of the Czech Republic* (Prague: MoFA Press).

Ministry of Industry and Trade/Czech Statistical Trade Office (2009) *Foreign Trade of the Czech Republic 2008* (Prague: MoTI Press).

Perthes, V. (1998) Germany gradually becoming a Mediterranean state, *EuroMeSCo Paper* 1, Lisbon.

Polish Ministry of Foreign Affairs (2003) *Strategia Polskiej Wspólpracy na rzecz rozwoju przyjeta przez Rade Ministrów w dniu 21 pa'zdziernika 2003 r* (Warsaw: MoFA Press).

Prizell, I., Nitze, P. H. & Michta, A. A. (Eds) (1995) *Polish Foreign Policy Reconsidered. The Dilemmas of Independence* (Basingstoke: Palgrave).

Rozsa, E. (2005) National attitudes of new EU member states towards the EMP: the case of Hungary, *EuroMeSCo Paper* 42, Lisbon.

Rozsa, E. (2009) New EU member states' positions regarding the Union for the Mediterranean. The case of Hungary (unpublished manuscript).

Schmid, D. (2008) Die Mittelmeerunion – ein neuer französischer Motor für die europäische Mittelmeer-Politik? *DGAPanalyse Frankreich,* January. Deutschen Gessellschaft für Auswärtige Politik, Available at http://www.zukunftsdialog.eu/fileadmin/user_upload/Bilder/Publikationstitel/2008_01_dgapana_f_schmid_www-2.pdf

Schröder, G. (2006) *Entscheidungen. Mein Leben in der Politik* (Hamburg: Hoffmann & Campe).

Schumacher, T. (2005) *Die Europäische Union als internationaler Akteur im südlichen Mittelmeerraum. 'Actor Capability' und EU-Mittelmeerpolitik* (Baden-Baden: NOMOS).

Schumacher, T. (2008) The German EU Presidency and the Southern Mediterranean, *EuroMeSCo e-news*, 11.
Schumacher, T. (2009) Germany: a player in the Mediterranean, in: IEMed/CIDOB (Eds) *Med.2009. 2008 in the Euro-Mediterranean Space* (Barcelona: IEMed/CIDOB).
Shapovalova, N. (2009) The EU's Eastern Partnership: still-born?, *FRIDE Policy Brief* 11, Madrid.
Webber, D. (2001) Introduction: German European and foreign policy before and after unification, in: D. Webber (Ed.) *New Europe, New Germany, Old Foreign Policy? German Foreign Policy since Unification* (London: Frank Cass).
Wojna, B. (2008) Poland and the Mediterranean (unpublished manuscript).
Zaidi, S. & Sami, D. (2009) Czechs seek role as honest broker in the Middle East, *Prague Wanderer*, 25 February.

# The UfM and the Middle East 'Peace Process': An Unhappy Symbiosis

ROSEMARY HOLLIS
Professor of Middle East Policy Studies, City University, London

ABSTRACT  *This contribution explores differing theories on how the failure of the 'peace process' featured in the design and goals of the UfM, drawing on lessons from the period when the EMP was pursued in parallel with the peace process. In each case, institutional overlaps are identified, as well as commonalities in the approaches of the actors to both pursuits. Crucially, however, the persistence and intensification of the Arab–Israeli conflict, in combination with the shift from multilateralism to bilateralism embodied in the UfM, has politicized the latter at the expense of the functionalist aspirations of its architects.*

The objective here is to define the relationship between the Union for the Mediterranean (UfM) and the unsuccessful Middle East Peace Process (MEPP). The operating assumption is that the Arab–Israeli conflict represents a key feature of the context within which the UfM was launched and one of the main questions explored is whether the UfM was created as a way to downscale European ambitions in the face of a deteriorating situation in the Middle East. Then, in keeping with the line adopted in the framework paper for this collection, the argument developed is that the UfM could not avoid entanglement with the conflict.

Contrasting views on the impetus behind the UfM are examined in the first section below and, as will be seen, the verdict on how it relates to the MEPP depends in part on how one understands the relationship between the Euro-Mediterranean Partnership (EMP) and the MEPP. As discussed in the second section, the vision embodied in the EMP could not be realized without a resolution of the Arab–Israeli conflict and the architects of the EMP tended to assume that the MEPP would take care of that.

On this they were disappointed, but the failure was due more to flaws in the design of the peace process in the 1990s than to the weaknesses of the EMP. Also, while the failure of the MEPP was a blow to the EMP, this was only one of the factors that led to disillusion and frustration with the EMP and thence the creation of the UfM.

As argued in the third section, however, in so far as the UfM did represent an attempt to refocus European ambitions in the face of reversals in the peace process, its fate will rest on whether it can be ring-fenced or insulated from the Arab–Israeli conflict. The likelihood of achieving such a separation is deemed remote, principally because of changes in the regional and international context stemming from 'the war on terror' declared after 11 September 2001 (9/11) and the fallout from the invasion of Iraq. As a result of these developments, the salience of the Israeli–Palestinian conflict has increased, rendering it less plausible that the UfM can attain its modest objectives in isolation.

There are other reasons why the fate of the UfM is inseparable from that of the dysfunctional MEPP. The UfM has been built on the edifice of the EMP and this was informed by assumptions about the normative mission of the European Union. Thus too much hope has been invested in the capacity of so-called 'soft power'[1] to effect change in the Mediterranean Partner Countries (MPCs), notwithstanding persistent failures by the Europeans to act on or abide by their professed norms. As argued in the fourth section, European declaratory policy represents a retreat from reality into the safety of simple pronouncements and self-serving projects.

In the fifth section attention turns to some of the individual European and Arab actors and their stance on the MEPP. As demonstrated, even if the Europeans put risk avoidance above conflict resolution, for the Arab states the need to keep conflict resolution on the international agenda, including through the UfM, is a matter of survival.

Another reason for scepticism that the UfM can succeed in attaining its limited goals has to do with the institutional framework within which the EMP and now the UfM, as well as the MEPP operate. The MEPP is an international endeavour, led by the United States, and the UfM is a regional initiative, presided over by a joint presidency (starting with France and Egypt). Yet the two are institutionally linked through a plethora of mechanisms, including the Middle East Quartet (that groups the EU, UN, US and Russia), the Arab Peace Initiative (devised and agreed by the Arab League, which now participates in the UfM) and the European Neighbourhood Policy (ENP). The implications for the UfM of these overlapping institutional arrangements are explored in the sixth and final section of this contribution.

**The UfM and the Conflict: Adapting to Realities?**

As Bechev and Nicolaidis (2008) and Kausch and Youngs (2009) have suggested, the UfM was conceived partly as a way to forge closer economic ties and security co-operation between the northern and southern shores of the Mediterranean *notwithstanding* the conflict. Although these commentators are sceptical about the prospects, they detect a perception, among some of the original architects of the scheme, that the conflict could be sidelined or 'parked' as a separate concern, to be dealt with primarily within the framework of the MEPP. On the latter, in contrast to the UfM, the United States has been expected to take the lead.

As explored in other contributions to this collection, the motives of the individual European players who signed up to the UfM were varied. For France the initiative represented an opportunity for President Nicolas Sarkozy to take a lead on matters Mediterranean (see Delgado and Gillespie, this collection). The French also identified a need to make up for the flagging fortunes of the EMP. Spain scrambled to respond to the challenge the French lead posed to its role in the Mediterranean (see Gillespie, this collection). Germany was initially critical of the initiative because Sarkozy appeared set on pursuing his scheme at the expense of the EMP (see Schumacher, this collection). Turkey, in turn, reacted negatively at first for fear that the proposal would affect its bid to join the EU (Balfour and Schmid, 2008; Emerson, 2008).

In all these cases, the capacity of the UfM to affect or be affected by the quest for Arab–Israeli peace seems not to have featured centrally in the actors' calculations. By contrast, Israel's initial response to the UfM was influenced by concerns that the Arab League would be accorded a bigger role in the plan than it had been given in the EMP (see Bicchi, this collection; Del Sarto, this collection). In other words, the Israelis feared spillover from the conflict into the proceedings of the UfM. For some Arab states, meanwhile, there could be no question of forgoing any new opportunity to air their frustrations with Israel's continued occupation of Arab lands captured in 1967. Arab resistance to 'normalizing' their relations with Israel, following the Israeli assault on Gaza in 2008–09 and the advent of a more hardline Israeli government since then, have already undermined the progress of the UfM (see in particular Schlumberge, this collection; Barber, 2010; Vogel, 2010).

Turning to the official pronouncements of the UfM, in these the desirability of conflict resolution is acknowledged but not emphasized. At the launch in Paris in 2008 the heads of state and government reaffirmed their 'support for the Israeli–Palestinian Peace Process'; recalled that peace in the Middle East 'requires a comprehensive solution'; and welcomed 'the announcement that Syria and Israel have initiated indirect peace talks under the auspices of Turkey' (Joint Declaration, 2008: Article 7). Yet, as the EMP before it, the UfM espouses the objective of turning the Mediterranean basin into an area of peace, stability and prosperity through dialogue and co-operation (Barcelona Declaration, 1995; Joint Declaration, 2008). Arab–Israeli peace is thus *ipso facto* incorporated into the vision of the UfM.

The linkage between the UfM and the MEPP is also evident in the orientations, calculations and policies of European and Arab actors and Israel in their dealings with each other.[2] In institutional terms there are also overlaps and interconnections, as discussed below. The thrust of this inquiry, therefore, is to determine how closely the fate of the UfM is tied to that of the dysfunctional MEPP.

This question is clearly not quite the same as asking what effect the continuance of the Arab–Israeli conflict will have on the prospects for the UfM. If the conflict continues, by definition the goal of developing an area of peace in the Mediterranean will not be met. Yet that does not eliminate the possibility that the UfM could pave the way for conflict resolution by generating more co-operation in other spheres, such as economic development, job creation, intelligence sharing and cross-cultural understanding (Aliboni and Ammor, 2009).

What matters, therefore, is the underlying logic with which the various stakeholders approach both the UfM and the MEPP. Belief in the power of economic development and institutional reform to counter instability and conflict informed the EU approach to *both* the EMP and the MEPP, but in neither case did the Europeans live up to the principles they espoused (Tocci, 2005; Pace, 2009; Al-Fattal, 2010). On the Arab side, at least at the government level, a determination to retain power has infused their approaches to both the EMP and the MEPP (Fernández and Youngs, 2005; Pace, 2010). For Israel, the UfM represented a potential opportunity to subsume the conflict.[3]

The UfM largely dispenses with the normative agenda that characterized EU aspirations for the EMP in its early years (Aliboni and Ammor, 2009; Kausch and Youngs, 2009). However, that agenda was abandoned or reneged upon long before the launch of the UfM, notwithstanding persistent references to European 'values' in EU rhetoric.[4] The way political reform featured in the EU approach to the MEPP also represented a betrayal of the ideals purportedly espoused by the EU. This became apparent after the outbreak of the second Intifada in 2000 and more markedly following the election victory of Hamas in 2006, as discussed below.

Thus, even though the UfM has been greeted by some as a sign that realism has replaced wishful thinking, there is still room to question that conclusion. If the EU had truly woken up to reality, then its approaches to both the UfM and the MEPP should embody greater recognition of the growing dangers posed by the conflict.

Latterly, not only has the problem of extremist movements and transnational terrorism become a shared concern for European and Arab governments, but it has become conflated with the Arab–Israeli conflict. In addition, that conflict has worsened, compounded by the failures of the MEPP. The UfM does not offer a resolution and the elevation of North–South dialogue to state level increases the likelihood that the initiative will become hostage to the conflict.

## The EMP and the MEPP: Pointers for the UfM

The existence of a seemingly promising MEPP at the time the EMP was launched in 1995 enabled the latter to go ahead on the assumption that its vision for an area of peace, stability and prosperity could be realized (Dosenrode and Stubkjaer, 2002). It also made it possible to argue that the EMP would serve to 'underpin' the MEPP and anticipated peace deals (Marks, 1996: 2). However, the EMP was not specifically envisaged as an alternative or rival to the MEPP.

On the contrary, in the mid-1990s the preoccupation in Europe was not with the Arab–Israeli conflict but with the security problems posed by Islamist extremism emanating from North Africa and manifest in the bomb attacks in France in 1995.

The security agenda that the EMP was supposed to address had to do, first and foremost, with migration, arms proliferation, Islamist terrorism, instability and economic malaise in the south (Bicchi, 2007; Hollis, 2000: 125). As argued in a report issued by the EU Institute for Security Studies, one of the faults of the EMP was that it failed to face up to realities, such as the Arab–Israeli conflict (Aliboni *et al.*, 2008).

Over time a number of scholars and commentators have examined the achievements and failings of the EMP and by all accounts it has fallen short of meeting its declared goals. These included: a political and security partnership, establishing 'a common area of peace and stability'; economic and financial partnership, 'creating an area of shared prosperity'; and partnership in social, cultural and human affairs (Barcelona Declaration, 1995). It is on the second of these objectives that most attention has focused and the verdict commonly reached is that prosperity has eluded the majority of people in the MPCs and the gap between standards of living in the north and south has widened (Joffé, 1999; Radwan and Reiffers, 2005). Economic growth and development in Israel, by contrast, has overtaken that of some EU countries (Nathanson and Stetter, 2005).

Assessments of the progress of the EMP in achieving its other objectives have been largely scathing. Pace (2007), among others, deconstructs the normative approach of the EU and documents the pernicious consequences for democracy promotion in the MPCs. Al-Fattal (2010), focusing on EU aid and policies to build a functioning democracy in the Occupied Palestinian Territories (OPT), charges the EU with reneging on its own principles and promises, to the detriment of the Palestinian people and their prospects of statehood. Pace (2007), Al-Fattal (2010) and Le More (2005) highlight the failure of the EU to tackle the core problem facing the Palestinians, namely: the Israeli occupation.

With respect to the security agenda, in an early assessment Spencer (1997) contends that the EU proved unable to develop a partnership with MPCs distinctive from or co-ordinated with other initiatives such as those of the North Atlantic Treaty Organisation (NATO) and the Western European Union (WEU). She also points to the contradictions inherent in the EU approach to Mediterranean security, in so far as it hoped to protect Europe from migration and instability in the South, while speaking the language of partnership. Aliboni and Ammor (2009), among others, highlight the betrayal of the EMP vision implicit in the increasing emphasis placed on intelligence sharing and security co-operation between governments, at the expense of political reform.

A common feature of many assessments of the EMP, shared by this author, is that it failed to meet expectations largely because the Europeans hoped to use it as a vehicle for the export of values and practices that could not and would not meet with acceptance in the MPCs. Conceivably, therefore, the Europeans were either unrealistic or naïve, or a combination of both. A close reading of EU pronouncements and stated expectations gives credence to the latter.[5] The alternative explanation, namely that the Europeans were disingenuous, would mean that while they claimed to want to promote democracy, free trade and co-operation, they actually intended to use such claims as a cloak for furthering their own interests, including stemming the flow of migration from South to North. To hold up, this explanation would imply an absence of the kind of assumptions about the benefits of adopting EU values that informed the whole EU enterprise (Le More, 2005; Hollis, 2009).

Irrespective of whether the Europeans were naive or disingenuous, it may be that the EU, weakened by the competing interests of member states, was outmanoeuvred by Arab governments and Arab elites intent upon protecting their interests (Hamzawy, 2005; Hollis, 2009; Khouri, 2010). Equally important, as argued here, is that intervening events, in particular 9/11 and the issue of terrorism linked to Islamist groups, transformed priorities in Europe and in the MPCs, overriding the reform agenda (Fernández and Youngs, 2005).

Turning to the MEPP, it could conceivably be argued that its failure was either a symptom or a consequence of the flaws in the EMP. However, neither the EMP nor the MEPP can be reduced to a dependent variable of the other. The two endeavours coexisted and the verdict here is that the MEPP pursued in the 1990s, under US leadership, *was itself flawed* (Keating et al., 2005; Miller, 2008) and only one of its shortcomings could be laid at the door of the EU for focusing on state-building in the OPT to the neglect of policies designed to bring an end to the occupation. After the outbreak of the second Intifada the EU made more mistakes, as discussed below. The key point here, however, is that, in the 1990s at least, the failings of the EMP cannot be attributed solely or even mostly to the dysfunctional MEPP per se.

The context changed after 2000 and the collapse of the MEPP, partly as a result of the second Intifada and changes in government in the United States and Israel, but also because of 9/11 and all that flowed from that. How these developments affected European and Arab attitudes towards the MEPP and the EMP is the subject of the next section.

## The UfM in Context: No Escape from the Conflict

At various stages in the past 40 years the Europeans, the Maghreb and the Arab Gulf states have proceeded on the assumption that progress could and should be made toward closer economic ties and security co-operation irrespective of the continuance of the Arab–Israeli conflict. During the 1970s and 1980s, for example, the Europeans developed their relations with the Arab world and Israel on a bilateral basis (Ismael, 1986) in a manner deliberately intended to ensure that the continuance of the conflict would not be allowed to interfere with their pursuit of closer commercial links and energy security.

In the 1990s by contrast, complementarity and convergence were assumed and welcomed between the EMP and the MEPP. When the latter collapsed there was a reversion to bilateralism, including at the level of EU–Arab and EU–Israeli relations, through the ENP (see Bicchi, this collection). However, de-linkage between the conflict and schemes for Mediterranean integration was no longer an option. The US declaration of 'the war on terror' following 9/11 and the disruption of the regional balance of power resulting from the invasion of Iraq in 2003 saw to that.

The 'war on terror' meant that all US allies were expected to demonstrate their loyalty: 'You are either with us or you are with the terrorists!' The reaction in Europe was initially strongly supportive of the United States in its hour of need. French President Jacques Chirac and British prime minister Tony Blair went to Washington to signal solidarity. Blair declared that Britain stood 'shoulder to shoulder' with America.[6] NATO invoked the Treaty to support military action against al-Qaeda in Afghanistan. However, in a portent of what was to come, Washington declined to make use of NATO in the initial stages of the subsequent invasion. Having learned from experience in the Balkans, the Americans were in no mood for the frustrations of coalition warfare.

Serious divisions emerged in the transatlantic alliance over how to combat al-Qaeda-inspired terrorism (Gordon and Shapiro, 2004). By its very nature, the EU is committed to internationalism and international law. The Bush administration, by contrast, positively relished the prospect of acting alone, informing the Europeans that their job was to do 'the nation-building' in the wake of US war-fighting (Kagan, 2002; Gordon and Shapiro, 2004).

In the background, under the leadership of Ariel Sharon, the Israelis depicted themselves as steadfast allies fighting 'in the same trench' as the Americans against the menace of Islamist-inspired terrorism. Palestinian President Yasser Arafat was denigrated as a sponsor of terrorism and blockaded in his compound in Ramallah. European governments refused to go along with the delegitimization of Arafat and their officials continued to pay him visits, but they failed to convince the Americans or the Israelis to take him seriously any more as a 'partner for peace'. EU policy became focused on the limited objective of keeping the Palestinian Authority (PA) afloat in the face of a crushing Israeli assault on its operations and infrastructure (Hollis, 2004).

After the fall of the Taliban regime in Kabul in late 2001 it soon became apparent that the United States planned to take the war against terror into the Middle East. In his State of the Union address in January 2002 Bush designated Iran and Iraq along with North Korea as part of 'an axis of evil'. For the remainder of the year the prospect of a US invasion of Iraq gathered pace and the EU divided.

A semblance of unity was restored when the UN Security Council adopted Resolution 1441, initiating a new round of weapons inspections in Iraq. However, when the inspectors failed to find conclusive evidence of an active Iraqi weapons programme, the Security Council split. France called for more time for the inspectors. The British joined the Americans in frantic lobbying at the UN, but failed to gain a green light for an invasion. Blair nonetheless convinced the British Parliament to agree to military action, and in March 2003 British troops accompanied the Americans into Iraq. Because the Turkish Parliament refused to facilitate entry from the north, Kuwait provided the sole point of access for the ground assault.

Once Saddam Hussein's regime was toppled and the UN formally recognized the occupation of Iraq, Washington called on its allies to help quell the resistance. The East Europeans offered the most enthusiastic support. Spanish troops were also deployed, but withdrawn again, in the wake of the Madrid bombings. As these demonstrated, far from reducing the threat of Islamist-inspired terrorism, the invasion of Iraq exacerbated it. Meanwhile, both Arab and European allies of America tried to persuade Washington that the plight of the Palestinians was part of the problem. The United States was not convinced.

Offering a ray of hope, in 2002 Bush did issue the first formal US endorsement of a two-state solution to the Israeli–Palestinian conflict. Seizing on this opportunity, the EU set about drafting the roadmap that they hoped would turn Bush's 'vision' into reality. Yet Sharon outmanoeuvred them, conditioning Israeli co-operation on the marginalization of Arafat and securing US recognition that Israel could not be expected to return to the 1967 borders in the event of a peace deal.

The lesson here is that the Europeans, with or without Arab concurrence, were powerless to push the MEPP in the absence of US and Israeli commitment to a solution. Added to which, because of differences among the Europeans over the Iraq crisis, the ambition of the EU to achieve a unified and effective foreign policy stance suffered a severe setback. On the need for resolution of the Arab–Israeli conflict they could still agree, but divisions over Iraq reduced European leverage and credibility in Washington. As US forces faced a full-scale insurgency and sectarian conflict in Iraq, European opponents of the invasion forbore to say 'I told you so!' Instead, all made efforts to repair their relations with Washington and in the process re-invigorated their commitments to combating terrorism through intelligence co-operation and new strictures on their own migrant populations and asylum seekers.

Among these measures, in 2003 the EU decided to add Hamas to its list of terrorist organizations (Hroub, 2006: 113–16). This move may have helped marginally to appease Washington and the Israelis, but also paved the way for EU paralysis after Hamas won the Palestinian legislative elections in January 2006. EU commitment to pursuing the MEPP through the Middle East Quartet also served to dilute the role of the EU and signalled a new preoccupation with policy co-ordination across the international community as opposed to effective action, as discussed below.

The fallout from the invasion of Iraq meanwhile transformed the context of the MEPP. Two consequences deserve mention. First, having found none of the weapons programmes cited to justify the invasion, the Bush administration espoused a new rationale, namely regime change and democratization across the region. As the United States launched its reform initiative for the Broader Middle East and North Africa in June 2004, the EU responded with its own strategy, namely the ENP, also launched in 2004. Having invested so much through the EMP in promoting economic and political reform in the MPCs, the EU hoped to demonstrate more substantive results through differentiation and Action Plans tailored to the specific needs and capacities of the partner countries. In the process, an opportunity to build in conditionality related to conflict resolution was overlooked.

Secondly, the invasion of Iraq produced a surge in anti-Americanism (Center for Strategic Studies, 2005), not only in the region but among Muslim populations everywhere, including Europe. Resentment of US high-handedness proved more widespread, in conjunction with a rise in Islamophobia. The principal beneficiary in the region was Iran and its allies Hizballah in Lebanon and Hamas in the OPT. Hizballah soon became embroiled in war with Israel in July 2006. The psychological victory was claimed by Hizballah, since its forces managed to keep up a hail of rockets into Israel until the day a ceasefire was finally agreed.

During the 2006 Lebanon war the Europeans and Arabs were again divided (Hollis, 2010). The French were among those calling for an immediate ceasefire. Britain held off doing so immediately in the hopes that Hizballah could be quelled first. Saudi Arabia and Egypt also hoped that Hizballah would be chastened, but ended up having to join the general Arab outcry against the Israeli bombardment of Lebanon.

Within two years Israel went to war again, this time against Hamas in Gaza. The Palestinians lost over 1,000 lives, many of them women and children, while Israeli dead numbered only 13. In Europe, the fate of the Palestinians in the Gaza war of 2008–09 gained them increased popular support. Yet when Israel comes in for criticism for its treatment of the Palestinians, there is always the danger that this could conflate with anti-Semitism. Certainly the Israelis are attentive to this possibility and have used the spectre as leverage on European governments.

As the foregoing demonstrates, thanks to the fallout from the invasion of Iraq, the Middle East has become more unstable. The old order, dominated by the Arab states, has been weakened and the new beneficiaries are the non-state actors Hizballah and Hamas, together with Iran.

In this situation all the stakeholders in the Arab–Israeli conflict cannot consider it a localized problem, capable of marginalization. Since 2009 the Syrians have gained new prominence as the potential weak link in the Iran–Syria–Hizballah axis. President Assad was feted by President Sarkozy as the star at the launch of the UfM, simply for turning up, and France has since proceeded with upgrading relations with Damascus without the latter having to change any aspect of its regional posture or domestic politics (ICG, 2009). Damascus has also been the object of US diplomatic approaches since Obama came to power. Yet without Israeli co-operation the Americans cannot wean the Syrians away from their alliance with Iran or their support for Hizballah and Hamas.

France may have wanted the dual presidency of the UfM, and selected Egypt to serve alongside France in the first instance, for reasons unrelated to the Arab–Israeli conflict. Egypt, by contrast, could not miss an opportunity to raise the issue. As the leading state representative of the Arabs in the UfM, in which the Arab League is also represented and could upstage Cairo, Egypt's regional standing and prestige are at stake. Turkey's new profile in the region, its defence of the Palestinians at the time of the Gaza war and its reaction to Israel's commando raid on the flotilla that sought to break the Gaza blockade in 2010 represent a challenge to Egypt. In short, if ever the possibility of sidelining the Arab–Israeli conflict existed, it is no longer feasible.

**Europe's Misleading Normative Narrative**

Attention now turns to the evolution of European thinking on the Arab–Israeli conflict. European perspectives have progressed through several phases since the British and French finally exited their last imperial domains. In the 1970s the first oil price shock and fear of Arab reprisals for support for Israel persuaded most Europeans to embrace a more pro-Arab stance on the conflict (Ismail, 1986). By the time the Europeans chose the Arab–Israeli conflict as the test case for the development of a common European foreign policy, culminating in the Venice Declaration (1980), the United States had taken the lead in conflict mediation.

Israel rejected the Venice Declaration and Washington proceeded with its own diplomacy. After Israel invaded Lebanon in 1982, France attempted its own mediation policy, alongside the United States, including deploying troops. Yet both their efforts foundered. After the leadership of the PLO was evacuated to Tunis, the Europeans proved unable to adopt a unified approach to dealing with the organization. However, the scene changed in 1990–91 as the Cold War ended and the United States marshalled an international and Arab coalition to reverse the Iraqi invasion of Kuwait. When the United States and post-Soviet Russia convened the Arab–Israeli peace conference in Madrid, the Europeans were accorded observer status only.

Since then and substantively in the 1990s the EU has played junior partner to the United States in the pursuit of conflict resolution (Dosenrode and Stubkjaer, 2002). After Norway brokered the so-called Oslo Accords, Washington took charge of driving what became the MEPP. The EU participated in the multilateral process that accompanied bilateral negotiations until the mid-1990s (Peters, 1996) and presided over the Regional Economic and Development Working Group (REDWG). The EU became the single largest donor to the Palestinian Authority, elected and constituted under the MEPP (Al-Fattal, 2010: 51–3).

In the 1990s EU policy on the Mediterranean and the MEPP could be characterized as a 'soft-power' approach to addressing European security needs (from migration to terrorism) through the disbursement of economic development aid, institution building and the promotion of good governance. Thus, when the PA was established in the OPT the Europeans, individually and collectively through the EU, concentrated on giving development aid and helping the PA govern the Palestinians. This suited the Americans, who reserved management of the political negotiations between the Palestinian leadership and Israel to themselves. In both respects there was palpable progress initially, but after the election of a Likud government led by Netanyahu, he proved resistant to implementing the provisions of Oslo.

Meanwhile, the expansion of Jewish settlements in the OPT continued. Periodic terrorist attacks on Israelis by Palestinians, some under the Hamas banner, also undermined confidence. However, given the devotion of the Clinton administration to pursuing peace and renewed hope following Labour's return to power in Israel in 1999, the Europeans apparently saw little purpose in breaking ranks with the Americans. However, retrospective analysis on the MEPP during the 1990s, has deemed it flawed (Keating *et al.*, 2005). Not only did it require the Palestinians to build a state-in-waiting while still under occupation and with no guarantee from Israel or even Washington at that stage that statehood would be the reward, but US diplomacy proved partial to the Israelis (Miller, 2008).

In effect, the Europeans became complicit in a US-led strategy that required the Palestinians to control their own militants and forgo resistance in the name of convincing the Israelis to end the occupation. Yet, in so far as the PA did so, the Israelis could sustain the occupation. In terms of the dichotomy between functionalism and politicization depicted by Bicchi in the framework for this collection, with respect to the MEPP the EU has opted for a functionalist approach to Palestinian state-building which has turned out to have highly political consequences.

Rather than draw attention to the occupation and focus on bringing it to an end, the EU may have actually helped to perpetuate it, on which, see more below. In any case, when the second Palestinian Intifada erupted in 2000, the peace process collapsed and the Israelis responded by reasserting control by force. The PA only survived thereafter thanks to EU support.

The EU was by this time a member of the Quartet and through this mechanism worked to develop the roadmap that was supposed to turn into reality what Bush announced as his 'vision' for a two-state solution in 2002. Thereafter, policy co-ordination, within Europe and across the Quartet, substituted for action. Inability to force through implementation of the roadmap symbolized the ineffectiveness of the MEPP in the first years of this century. This failure turned to counterproductive meddling after the Hamas victory in the Palestinian legislative elections of January 2006.

During and following those elections the United States briefed members of the Palestinian Fatah movement to resist any form of compromise or co-operation with their Hamas rivals for power in the PA. The Quartet formulated three principles that Hamas was expected to embrace to gain acceptance: renunciation of violence; recognition of Israel's right to exist; and acceptance of all agreements previously signed between the Palestinian leadership and Israel. This was a recipe for stalemate. When Saudi Arabia brokered a power-sharing agreement between Fatah and Hamas in 2007, Washington scuppered the deal.

*Realism or Retreat?*

From the Venice Declaration of 1980 to the Conclusions of the Council of Ministers in December 2009, the EU has led the way on declaratory policy. In their conclusions, the ministers called for 'a two-state solution' to the Israeli–Palestinian conflict, with 'the State of Israel and an independent, democratic, contiguous and viable State of Palestine, living side by side in peace and security' (Council of the EU, 2009).

The Council also stated that the EU 'will not recognize any changes to the pre-1967 borders including with regard to Jerusalem, other than those agreed by the parties' and that, in the interests of 'genuine peace, a way must be found through negotiations to resolve the status of Jerusalem as the future capital of two states'. Beyond this, the Council noted that: 'A comprehensive peace must include a settlement between Israel and Syria and Lebanon.'

A declaration of what should happen is not, however, a plan of action. Successive EU statements have not spelled out what the Europeans would do to make the parties to the conflict conform to their wishes. Also, while pronouncing on what is required for an end to the conflict, EU dealings with the protagonists have not made progress on bilateral relations conditional upon the implementation of steps to reach a two-state solution. Israel has not been punished for continuing the occupation, house demolitions, the confiscation of Palestinian homes and land, or construction of the security barrier that was deemed in contravention of international law by the International Court of Justice in 2004.

The Palestinians in the Gaza Strip have languished under a blockade that the EU has branded unacceptable and counterproductive. What finally prompted Israel to ease the blockade – but not end it – in June 2010, was not EU action but the fallout from the deadly Israeli commando raid on the flotilla of ships mounted by civilian volunteers that challenged the blockade. Turkey, not the EU, applied most pressure and Washington called for a re-think (BBC, 2010; Khalaf, 2009).

Meanwhile, the EU has been active in supporting Palestinian institution-building under the emergency administration of prime minister Salam Fayyad in the West Bank. Europeans are working with the Americans and Canadians to train Palestinian police and security forces to keep law and order in Palestinian towns (Asseburg, 2009b). The logic of these endeavours rests on the assumption that the PA must be prepared to take on the responsibilities and tasks of government when statehood is realized. The EU pays the salaries of PA police, civil servants, teachers and medical workers, including those of Fatah members in the Gaza Strip who are instructed to stay at home rather than work for Hamas-run organizations. According to European Commission figures, between 2000 and 2009 the EU disbursed over €3.3bn in aid to the Palestinians.[7] Al-Fattal (2010), Brown (2010) and Pace (2010) argue that this approach has been counterproductive and a disservice to the Palestinians.

The net result of EU endeavours is conflict management, not resolution. In contrast to the situation in the 1990s, when European aid was channelled into development projects in the West Bank and Gaza, today the EU only finances the running costs of a PA which is not subject to legislative oversight and whose remit only runs in so-called Areas A and B, while the remainder of the West Bank (60 per cent) is still under Israeli control and in which settlements have not been removed or curtailed.

The unelected PA owes its survival to the EU, but it only serves to keep the peace, improve internal governance and regulate business within Palestinian towns and villages which remain separated by Israeli checkpoints, interspersed with Israeli settlements and by-pass roads, sealed off from Israel by the security barrier and separated from Gaza.

Thus, the EU is not totally inactive. On the contrary, its members are busy and engaged, but only on the Palestinian side and this in ways that have undermined Palestinian unity and prospects. European support for the PA in the West Bank and complaints about the blockade of Gaza add up to no more than a holding strategy as opposed to a transformative one. EU tax payers' money is being used to relieve the Israelis of the costs of occupation and EU policy is not directed at rolling back that occupation.

## The Role of Individual Actors

Here attention turns to the positions of three European states (Britain, France and Germany) and two Arab states (Egypt and Jordan) on the MEPP. Whereas in the past each believed in their individual capacity to make a contribution to that process and invested resources accordingly, latterly they have retreated to risk avoidance in the case of the Europeans and survival mode in the case of Egypt and Jordan.

## Britain, France and Germany

The positions of all the European actors on the Middle East 'peace process' can be identified across a spectrum of positions on: (a) their relative sympathy or antipathy to the positions of the core protagonists in the conflict; (b) the importance they attach to their bilateral relations with Washington; and (c) the costs and benefits of raising their profile in the MEPP.

Taking as a benchmark the categorizations of different actors in the UfM introduced by Bicchi and discussed in other contributions here, over the past 20 years Britain, France and Germany have sampled the roles of 'entrepreneur', 'leader', 'veto-player', and 'low-profile supporter'. The descriptions 'favour exchangers' and 'unhappy laggards' do not fit. In all their roles, however, the three states have assumed that they are not acting in a vacuum and that the involvement of other Europeans and the United States is a pre-requisite for a successful peace process.

Whereas between 1948 and 1967 France was the leading supplier of arms to Israel, as of the 1967 war it decided against this role and turned its attention to developing better relations with the Arab world. Thereafter, until Sarkozy arrived on the scene, France could be identified as more sympathetic to Arab than Israeli concerns. Demonstrating independence from Washington also featured in French motivations (Hoffman, 1971), along with seizing opportunities to provide leadership in Europe.

In the 1980s France adopted the role of 'entrepreneur' in several contexts, notably in Lebanon following the Israeli invasion of 1982. France was also an open supporter of the Iraqi regime in the Iran–Iraq war, when Britain and the United States preferred to give limited support to Baghdad in secret and through Arab surrogates.

In the 1990s France championed the cause of expanding EU relations with the Maghreb, and through the EMP sought to match German initiatives for Eastern Europe. Itself the target of terrorist action linked to Algeria, Paris led on development of security co-operation between Arab governments within the context of the EMP. France was also far less enthusiastic about the political reform agenda than the northern Europeans (Youngs, 2006). With respect to the MEPP, France was in the forefront of European calls for attention to international law in the formulation of peace proposals, at the same time as developing projects in the OPT designed to garner publicity for France.

In contrast to France, since the Suez debacle the British have accorded high importance to maintaining close relations with Washington. Until the 1980s the British were also more sympathetic to the Arabs than Israel, but during the 1980s Margaret Thatcher took a tough line on terrorism, including refusing to meet members of the PLO. The arrival of Tony Blair at 10 Downing Street marked a new period of British activism. Blair played the role of would-be entrepreneur on several occasions, but always in a manner intended to help the US leadership in the MEPP (Hollis, 2010: 70–85, 135–57).

The Germans, by contrast to both the British and the French, have adopted the role of 'low-profile supporter' of the MEPP and occasionally 'veto player' within the

EU. Developing a close and supportive relationship with Israel was vital to ridding the Germans of the stigma of the Nazi era and making recompense to the Jews, through compensation payments to Israel (Lavy, 1996). Consequently, Germany could be counted on to veto any European initiative which could be depicted as biased against Israel. However, with the advent of the Oslo process, the Germans did begin to build a profile in the OPT with a diplomatic presence and assistance, including through the party Schtiftungen, to Palestinian projects and NGOs (see Schumacher, this collection).

Latterly, changes of government in Germany, France and Britain have led to shifts in their positions (see Schumacher, this collection; Asseburg, 2009a). Angela Merkel has gone out of her way to demonstrate German support for Israel, including speaking before the Israeli Knesset. Reportedly, one of her motives is to demonstrate that she, an East German, is prepared to own and atone for the Nazi past (Dempsey, 2010). Sarkozy has deliberately sought closer relations with Washington and shown greater warmth towards Israel than was typical of France in the past. Since the departure of Tony Blair from government in Britain, the British have declined to take a strong lead or even act as entrepreneurs in the context of the MEPP. Overall, given their preoccupations with adjusting to the new economic constraints affecting all three countries, it seems unlikely that the governments of Britain, France or Germany will be in the mood to launch any new initiatives on the Middle East for the foreseeable future.

*Egypt and Jordan*

Across the Arab world public support for the Palestinian cause and antipathy toward Israel is intense and volatile, As a result autocratic Arab regimes tend to use every opportunity to align themselves with such sentiments at the same time as trying to avoid having to take any actions that might turn animosity into war. Israel is not an opponent that any Arab state can contemplate engaging in battle without heavy penalty – as experienced by Lebanon in 2006 when Hizballah initiated a conflict that led to major destruction and over 1,000 Lebanese deaths, many of them civilians.

It is not surprising therefore that Egypt and Jordan, the only two Arab countries that have peace treaties with Israel, are constantly urging both the Americans and the Europeans to do more to end the Israeli occupation. In both cases the governments are frequently pressed by their publics to sever relations with Israel in protest at Israeli actions, but neither have done so, for fear of the consequences.

Herein lies one of the problems that befell both the US and the EU reform programmes. Real democracy in Egypt or Jordan could lead to the election of groups and parties that would scrap what they regard as their current governments' appeasement of Israel (see Schlumberger, this collection). It was for fear of this spectre and the rise of Islamist movements generally that the EU and the United States have ceased to press the reform agenda.

In addition, both Egypt and Jordan are in an especially difficult position because of their proximity to the Gaza Strip and the West Bank respectively. If Egypt opens its borders to Gaza and Jordan eases access for West Bank Palestinians, in both cases they would relieve Israel of some of the pressure to end the occupation.

As a consequence, both Egypt and Jordan lack the leverage to pressure Israel and must rely on others to do so. It is thus to be expected that both states will regard the UfM as an opportunity to make themselves heard. As mentioned above, as co-president of the UfM, Egypt in particular must regard it as a positive responsibility to keep the conflict on the agenda.[8]

## Institutional Overlaps and Conclusions

As discussed in other contributions here, it is possible to assess the institutional architecture of the UfM in terms of two dichotomies: regionalism versus bilateralism; and functionalism versus politicization. As the foregoing discussion indicates, in all respects the UfM is entangled with the institutional arrangements that frame the MEPP.

Unlike the UfM, the MEPP is an international endeavour and, as of 2002 and the formation of the Middle East Quartet, the United States, the UN, the EU and Russia have combined forces in pursuit of a common stance on Middle East peace. In 2002 the Arab League launched the Arab Peace Initiative (API), re-launched in 2007, as a collective Arab contribution to resolving the conflict. Within this context the EU and the Arab League, as well as some of the signatories to the API, are parties to the ENP and have signed up to the UfM.

To make clear distinctions between these structures, their functions and their goals is therefore unrealistic. To cite one example of functional overlap, EU engagement in the MEPP has been pursued through some of the same instruments and structures developed under the EMP and the ENP (Al-Fattal, 2010) which, in the latter case at least, continue to function. Thus the ENP Action Plan for the PA is actually more about fulfilling the expectations for Palestinian state-building envisaged as part of the MEPP than preparing the Palestinian economy for harmonization with the EU. In addition, successive EU collective and unilateral practical initiatives to aid Palestinian 'state-building' have embedded the European bureaucratic and security endeavours and personnel in the infrastructure of the occupation.

The linkages have become compounded over time. The Madrid process sought a comprehensive approach, but devolved onto bilateral tracks. Oslo was the central feature of this narrow approach, but when it failed, the effort to revive it became multilateral, through the Quartet. The API is a quest to shift from a bilateral approach to making peace deals (as was the case with Egypt and Jordan) to a regional or 'comprehensive' approach. Both the EU and the United States have sought, belatedly, to capitalize on this, but only in so far as the Arabs might be persuaded to 'deliver' the Palestinians to the table. The Arabs have refused to take any steps towards 'normalization' with Israel unless and until it withdraws from the Occupied Territories. This resistance to normalization is now being played out in the UfM.

Within this context, neither the UfM nor the unproductive MEPP is reducible to the status of a dependent variable in the relationship between the two. However, both are the product of the mindsets or worldviews of the actors involved, as repeatedly indicated above. In the case of the Europeans, they have operated on the assumption that the values embraced by all EU members and embodied in the *acquis* are not only positive for them but also for any other country (Hollis, 2009). Among the MPCs, meanwhile, a worldview prevails that is positively suspicious of EU intentions and values. These mindsets also informed the EMP and the MEPP in the 1990s, but have evolved in response to failures on both counts, as well as exogenous factors, including 9/11, the Iraq crisis, fallout from the invasion of Iraq in 2003 and the resulting intensification of the Arab–Israeli conflict.

The EU's declaratory policy on the MEPP, together with its narrow focus on Palestinian 'state-building' in the West Bank and the adoption of the limited objectives of the UfM indicate a retreat in the face of a gathering storm. For the Arabs, particularly Egypt and Jordan, that storm could spell destabilization. Hence the scene seems set for turbulence from which the UfM cannot be immune.

## Notes

[1] The term coined by Jospeh Nye and adopted by others, both academics and politicians, to contrast the EU approach to power projection with the military or 'hard power' available to the United States.

[2] As attested by officials participating in the seminar at which this and other papers were discussed in May 2010. As one said, official deliberations on the UfM and the MEPP are so interwoven as to be inseparable. If one tried to treat them as two separate clients for the purposes of billing for official time spent on each, the distinction drawn would be arbitrary or even false.

[3] An opportunity Israel apparently considered jeopardized by inclusion of the Arab League as a participant.

[4] For example, ahead of the Luxembourg summit in May 2005, Luxembourg's foreign minister Jean Asselborn declared that the EU was not just a source of funds but ought also to 'transfer European values to Arab society to encourage democracy' (Islam, 2005).

[5] Substantiated in interviews conducted by the author with EU officials in 2005–06.

[6] The role of Britain, and Blair in particular, is the subject of Hollis (2010).

[7] http://eeas.europa.eu/occupied_palestinian_territory/ec_assistance/eu_support_pa_2000_2009_en.pdf (accessed 22 November 2010).

[8] The problems connected with co-ownership are discussed by Johansson-Nogués (this collection).

## References

Al-Fattal, R. (2010) *European Union Foreign Policy in the Occupied Palestinian Territory* (Jerusalem: PASSIA).

Aliboni, R. & Ammor, F. (2009) Under the shadow of 'Barcelona': from the EMP to the Union for the Mediterranean, *EuroMeSCo Paper* 77, available at http://www.euromesco.net/index.php?option=com_content&task=view&id;=1142&Itemid=48&lang=en (accessed 31 January 2011).

Aliboni, R., Joffé, G., Lannon, E., Mahjoub, A., Saaf, A. & de Vasconcelos, A. (2008) *Union for the Mediterranean: Building on the Barcelona aquis* (Paris: EUISS).

Asseburg, M. (2009a) The Arab–Israeli conflict, in: *German Middle East and North Africa Policy*, SWP Research Paper 2009/RP 09, Berlin, September, available at http://www.swp-berlin.org/en/produkte/swp-studien/swp-studien-detail/article/german-middle-east-and-north-africa-policy-1.html (accessed 31 January 2011).

Asseburg, M. (2009b) The ESDP missions in the Palestinian Territories (EUPOL COPPS, EU BAM Rafah) peace through security? in: *The EU as a Strategic Actor in the Realm of Security and Defence?*, SWP Research Paper 2009/RP 14, Berlin, December, available at http://www.swp-

berlin.org/en/products/swp-research-paper/swp-research-paper-detail/article/the-eu-as-a-strategic-actor-in-the-realm-of-security-and-defence.html (accessed 31 January 2011) .
Balfour, R. & Schmid, D. (2008) Union for the Mediterranean, disunity for the EU? *European Policy Centre Policy Brief*, February.
Barber, T. (2010) EU's Union for the Mediterranean drifts into irrelevance, *ft.com/brusselsblog*, 1 June.
Barcelona Declaration (1995) Adopted at the Euro-Mediterranean Conference, 27–28 November.
BBC (2010) Gaza flotilla: Turkey threat to Israel ties over raid, 4 June. Available at http://www.bbc.co.uk/news/10236884 (accessed 31 January 2011).
Bechev, D. & Nicolaidis, K. (2008) The Union for the Mediterranean: a genuine breakthrough or more of the same? *The International Spectator*, 43(3), pp. 13–20.
Bicchi, F. (2007) *European Foreign Policy Making toward the Mediterranean* (New York: Palgrave).
Brown, N. (2010) Are Palestinians building a state? Available at http://www.carnegieendowment.org/publications/index.cfm?fa=view&id=41093 (accessed 22 November 2010).
Center for Strategic Studies (2005) Revisiting the Arab Street, report, Center for Strategic Studies, University of Jordan.
Council of the European Union (2009) Council conclusions on the Middle East Peace Process, 2985th Foreign Affairs Council meeting, Brussels, 8 December.
Dempsey, J. (2010) Embracing Israel costs Merkel clout, *International Herald Tribune*, 21 January.
Dosenrode, S. & Stubkjaer, A. (2002) *The European Union and the Middle East* (New York: Sheffield Academic Press).
Emerson, M. (2008) Making sense of Sarkozy's Union for the Mediterranean, *Centre for European Policy Studies Policy Brief*, 155, available at http://www.ceps.eu/book/making-sense-sarkozys-union-mediterranean (accessed 31 January 2011)
Fernández, H. A. & Youngs, R. (Eds) (2005) *The Euro-Mediterranean Partnership: Assessing the First Decade* (Barcelona: FRIDE).
Gordon, P. & Shapiro, J. (2004) *Allies at War: America, Europe and the Crisis over Iraq* (New York: McGraw Hill).
Hamzawy, A. (2005) Euro-Mediterranean Partnership and democratic reform in Egypt: contemporary policy debates, in: H. A. Fernández & R. Youngs (Eds) *The Euro-Mediterranean Partnership: Assessing the First Decade*, pp. 131–142 (Barcelona: FRIDE).
Hoffman, S. (1971) Franco–American differences on the Middle East, *Public Policy*, 14(4).
Hollis, R. (2000) Barcelona's first pillar: an appropriate concept for security relations?, in: S. Behrendt & C. P. Hanelt (Eds) *Bound to Cooperate: Europe and the Middle East* (Gütersloh: Bertelsmann Foundation).
Hollis, R. (2004) The Israeli–Palestinian road block: can Europeans make a difference? *International Affairs*, 80(2), pp. 191–201.
Hollis, R. (2009) European elites and the Middle East, in: A. Gamble & D. Lane (Eds) *The European Union and World Politics: Consensus and Division* (London: Palgrave Macmillan).
Hollis, R. (2010) *Britain and the Middle East in the 9/11 Era* (London: Wiley Blackwell and Chatham House).
Hroub, K. (2006) *Hamas* (London: Pluto Press).
ICG (International Crisis Group) (2009) Engaging Syria: lessons from the French experience, *Middle East Briefing*, 27, available at http://www.crisisgroup.org/en/regions/middle-east-north-africa/iraq-syria-lebanon/syria/B027-engaging-syria-lessons-from-the-french-experience.aspx (accessed 31 January 2011)
Islam, S. (2005) Another blueprint, *Middle East International*, 27 May, p. 22.
Ismael, T. (1986) *International Relations of the Contemporary Middle East* (Syracuse, NY: Syracuse University Press).
Joffé, G. (Ed.) (1999) *Perspectives on Development: The Euro-Mediterranean Partnership* (London: Frank Cass).
Joint Declaration (2008) Joint Declaration of the Paris Summit for the Mediterranean, Paris, 13 July.
Kagan, R. (2002) Power and weakness, *Policy Review*, 113, available at http://www.hoover.org/publications/policy-review/article/7107 (accessed 31 January 2011).

Kausch, K. & Youngs, R. (2009) The end of the 'Euro-Mediterranean vision', *International Affairs*, 85(5), pp. 963–975.

Keating, M., Le More, A. & Lowe, R. (Eds) (2005) *Aid, Diplomacy and Facts on the Ground: The Case of Palestine* (London: Royal Institute of International Affairs).

Khalaf, R. (2009) Ankara pursues central role in Changing Middle East order, *Financial Times*, 17 November.

Khouri, R. (2010) Arab civil society, limited and growing, *The Daily Star* (Beirut), 24 July.

Lavy, George (1996) *Germany and Israel: Moral Debt and National Interest* (London: Frank Cass).

Le More, A. (2005) Killing with kindness: funding the demise of a Palestinian state, *International Affairs*, 81(5), pp. 981–999.

Marks, J. (1996) High hopes and low motives: the new Euro-Mediterranean Partnership initiative, *Mediterranean Politics*, I(1), pp. 1–24.

Miller, A. D. (2008) *The Much Too Promised Land: America's Elusive Search for Arab–Israeli Peace* (New York: Bantam Books).

Nathanson, R. & Stetter, S. (Eds) (2005) *Israeli–European Policy Network Reader* (Tel Aviv: Freidrich-Ebert-Stiftung).

Pace, M. (2007) Norm shifting from EMP to ENP: the EU as a norm entrepreneur in the south? *Cambridge Review of International Affairs*, 20(4), pp. 659–675.

Pace, M. (2009) *Liberal or Social Democracy? Aspects of the EU's Democracy Promotion Agenda in the Middle East* (Stockholm: International Institute for Democracy and Electoral Assistance).

Pace, M. (2010) *Perceptions from Egypt and Palestine on the EU's Role and Impact on Democracy Building in the Middle East* (Stockholm: International Institute for Democracy and Electoral Assistance).

Peters, J. (1996) *Pathways to Peace. The Multilateral Arab–Israeli Peace Talks* (London: Royal Institute of International Affairs).

Radwan, S. & Reiffers, J. L. (2005) *The Euro-Mediterranean Partnership, 10 Years After Barcelona: Achievements and Perspectives* (Marseille: FEMISE).

Spencer, C. (1997) Building Confidence in the Mediterranean, *Mediterranean Politics*, 2(2), pp. 23–48.

Tocci, N. (2005) *The Widening Gap Between Rhetoric and Reality in EU Policy towards the Israeli–Palestinian Conflict*. CEPS Working Document no. 217, available at http://www.ceps.eu/book/widening-gap-between-rhetoric-and-reality-eu-policy-towards-israeli-palestinian-conflict (accessed 31 January 2011)

Venice Declaration (1980) *Bulletin of the EC*, 6-1980, pp. 10–11.

Vogel, T. Union for the Mediterranean summit postponed, *European Voice*, 25 May.

Youngs, R. (Ed.) (2006) *Survey of European Democracy Promotion Policies 2000–2006* (Madrid: FRIDE).

# *Plus ça change*...? Israel, the EU and the Union for the Mediterranean

RAFFAELLA A. DEL SARTO
OCHJS/Middle East Centre, St Antony's College, University of Oxford, UK

ABSTRACT *The patterns characterizing relations between Israel and the European Union comprise, firstly, repeatedly tense political ties that contrast with constantly deepening economic relations. Secondly the practice of bilateral relations markedly differs from their rhetoric. Thirdly, disagreements usually revolve around Middle East peacemaking. Finally, unlike the EU, Israel prefers disconnecting bilateral ties from regional politics. These patterns explain Israel's position and strategy toward the Union for the Mediterranean (UfM) and permit an assessment of the relevance of the latter for EU–Israeli relations. The conclusion is that the UfM is unlikely to alter the basic patterns of bilateral ties.*

The Union for the Mediterranean (UfM) was launched in July 2008 with the usual self-congratulatory excitement that has been characterizing EU foreign policy initiatives in recent years. As reported extensively, the initiative was a brainchild of the then candidate for the French Presidency, Nicolas Sarkozy, who had envisaged a policy initiative for Mediterranean states outside of the European Union framework. Yet, after some EU-internal haggling, the UfM eventually mutated into yet another EU-led attempt to 'reinvigorate' the comatose Euro-Mediterranean Partnership (EMP, or Barcelona Process) that had started in November 1995.[1] With a focus on rather technical projects in areas such as energy and transport, the initiative's inventors stressed the novel focus on institutionalization and co-ownership. Thus, in addition to establishing a co-presidency held by a 'northern' and a 'southern' head of state, the UfM will be governed by a secretariat based in Barcelona, in charge of implementing the projects.

The hype around the UfM was short-lived. While Israel's objection to the formal incorporation of the Arab League as an observer already resulted in a deadlock in August 2008, all UfM meetings were suspended at Egypt's request following Israel's offensive in the Hamas-ruled Gaza Strip the following December. For observers of Euro-Mediterranean affairs this development was hardly surprising

(Schumacher, 2009), as the EMP experience had amply demonstrated the relative ease with which regional initiatives came to depend on Middle East politics. With the second Netanyahu government in power in Jerusalem, there have also been growing political tensions between Israel and Brussels. Yet, given that Israel greatly appreciates the significant improvement of *bilateral* EU–Israel ties over recent years, and considering the country's aversion to EU-led *regional* projects, Israel's support for, and participation in, the emerging UfM deserves further investigation.

Indeed, why did the country not refuse outright to be part of the UfM? What role did Israel play behind the scenes vis-à-vis the emerging initiative? Did it act as 'leader', 'laggard', 'fence-sitter' (see Bicchi, this collection), or perhaps a mixture of them? What explains Israel's position? And what impact is the UfM likely to have on EU–Israeli relations? This contribution argues that in order to answer these questions, the main patterns characterizing EU–Israeli relations in the last two decades must be considered. It concludes that the UfM is unlikely to considerably alter the basic structure of EU–Israeli relations, which will remain instead *la même chose*.

## The EC/EU and Israel from the beginnings to Oslo

In the roughly 50 years[2] of official Euro-Israeli diplomatic relations, bilateral ties have developed slowly but constantly (Sachar, 1999). Due to European Jewish history, the State of Israel was initially not very keen on developing cordial ties with Europe. Israel and the then-European Communities (EC) signed the first economic agreement in 1964. Prior to the Six Day War of 1967, Israel felt abandoned by the Europeans because of the latter's diplomatic inactivity, with fears of a possible annihilation of the newly founded state mounting high among Israelis. Similarly, the EC's declaration after the 1973 war, which stressed legitimate Palestinian rights and the need for peace negotiations within the UN framework, was anathema to the government in Jerusalem. The latter also observed with consternation the launching of the Euro-Arab dialogue against the background of OPEC's threats to use the 'oil weapon' against Europe. Yet Brussels and Israel signed new trade agreements in 1970 and in 1975, i.e. shortly after the wars of 1967 and 1973 respectively, which entailed periods of strained relations.

But particularly the EC's Venice Declaration of June 1980 provoked the ire of the Israeli government. The declaration, it should be recalled, supported the Palestinians' right to self-determination and the association of the PLO with future negotiations, while stressing that the settlements in the occupied territories and any changes in the status of Jerusalem were illegal. Pointing to the terrorist nature of the PLO – at that time, it still advocated Israel's destruction – and referring to European Jewish history, Israel's government accused the Europeans of being willing to put Israel's security at risk for the sake of Arab oil supplies. In retrospect, the Venice Declaration was well ahead of its time. Yet, termed a 'defining moment in the Israeli discourse and in the public distrust of Europe' (Pardo and Peters, 2010: 9), the declaration was to set the tone for political relations for the years to come.

Indeed, in the following decades, variations of these accusations regularly followed the EC/EU's criticism of Israeli actions. Such occasions included Israel's 1982 invasion of Lebanon, its handling of the first Palestinian *intifada*, and the continuous expansion of Israeli settlements in the territories. Concurrently, the government in Jerusalem repeated its conviction that 'Europe' could not play any role in Middle East peacemaking as long as it embraced such ideas. Interestingly, former foreign minister Abba Eban made this point back in 1973 in response to the EC's declaration on the Yom Kippur war (Pardo and Peters, 2010: 7). The political disagreements also entailed that, notwithstanding Israeli requests, economic relations were not upgraded in that period.

This was to change with the beginning of the Oslo process in the early 1990s. The government under the late Yitzhak Rabin had requested European financial support of its peacemaking effort, and Brussels was happy to help (Sachar, 1999: 342; Del Sarto, 2006: 105–6). The beginning of the peace process also resulted in the EU's Essen Declaration of 1994 stipulating that Israel should enjoy a 'special status' in its relations with Brussels due to its advanced political and economic features (Council, 1994). This EU–Israeli honeymoon also resulted in the signing of an updated bilateral free trade agreement in 1995 against the backdrop of the EMP that had just been launched. Moreover, in 1996 Israel became the first (and so far only) non-European country to participate in the EU's so-called framework programme for research and development, a status it has maintained until the present. Yet relations were subsequently to become increasingly complex and contradictory, as discussed in the following section.

## Patterns of EU–Israeli Relations

In the last two decades, a number of patterns have come to characterize relations between the EU and Israel. The first pattern is the marked discrepancy between excellent economic ties, which are constantly deepening, and repeatedly tense political relations. As political disagreements tend to be voiced in public, there is, secondly, a wide gap between the practice of bilateral relations and their rhetoric. Thirdly, the disputes are usually related to the question of how to resolve the Arab–Israeli conflict, whereby right-wing governments in Jerusalem are more likely to clash with Brussels over this issue. Finally, the EU seeks to anchor its bilateral ties to Israel in regional politics, reflecting its ambition of playing a greater role in Middle East peacemaking. Conversely, Israel prefers disconnecting both tracks, aiming at a 'special relationship' with the EU in a sort of political vacuum instead.

*The Gap between Politics and Economics*

Disagreements in EU–Israeli relations returned the moment Israeli–Palestinian peacemaking entered an impasse, starting with the first premiership of Binyamin Netanyahu (1996–99). In this period, Brussels criticized the Israeli government for delaying the implementation of further Israeli withdrawals from the occupied territories according to the 1998 Wye River memorandum. The EU also started calling for an independent Palestinian state in its 1998 and 1999 declarations. But particularly its Berlin Declaration of March 1999, which referred to Jerusalem as a *corpus separatum* in accordance with the 1947 UN partition plan, prompted Israel's accusation that 'Europe, where one third of the Jewish population perished' endangered the State of Israel and undermined its interests (Ministry of Foreign Affairs of Israel, 2000).[3] The outbreak of the second *intifada* in 2000 – which entailed Israel's partial reoccupation of the West Bank from 2002 on – led to unprecedented tensions between both sides. While Brussels denounced Israel's military response, Israeli governments reacted with contempt to the EU's alleged incapacity to comprehend the need to fight terrorism. Profound differences also revolved around Israel's decision to treat the late Yassir Arafat as 'irrelevant' and to confine him to his *Muqata* headquarters, with European officials paying him visits by candlelight – in the absence of electricity. Moreover, the EU repeatedly criticized Israel's policies of targeted killings, closures, house demolitions, and settlement expansion. In return, Israel claimed that 'the Europeans' were biased and did not understand the country's security needs.

Bilateral relations improved with Ariel Sharon's Gaza disengagement plan, as it seemed to entail the resumption of peacemaking – erroneously, as it turned out. Brussels offered financial support for Israel's 2005 withdrawal from the Gaza Strip and also sent a border mission to the only crossing between the Strip and Egypt at Rafah. The Hamas victory in the Palestinian 2006 elections brought both sides closer, as Brussels aligned itself with the US in boycotting the Hamas-led government, much to Israel's delight. Yet, Israel's war on Hamas-ruled Gaza of December 2008 prompted a renewed exchange of accusations. Brussels repeatedly sought to pressure Israel to reach a permanent ceasefire and to lift its blockade on humanitarian aid in Gaza, imposed after Hamas seized power in June 2007 (Tocci, 2009). And Israel's raid on ships of a flotilla carrying aid and activists to Gaza at the end of May 2010, which left nine people dead, prompted EU foreign policy chief Catherine Ashton to demand an 'immediate, full, and impartial' inquiry (Council, 2010a: 1). Brussels also described Israel's blockade of the Gaza Strip as 'unacceptable and politically counterproductive' (Council, 2010b: 1).[4]

But whereas political relations were certainly not idyllic, economics presents a completely different picture. With an annual trade volume of around €25 billion, the EU remains Israel's largest trading partner. More than that, economic relations have improved significantly in recent years, following the launching of the European Neighbourhood Policy (ENP) in 2003–04. Indeed, due to its advanced economy, Israel was in a particularly good position to benefit from the tailor-made upgrading of bilateral relations that the ENP offered to its 'neighbours'.[5] Although close economic integration with the EU, even without membership, is economically advantageous for Israel (Tovias, 2006), Israeli governments were traditionally rather cautious vis-à-vis EU policy initiatives. Yet this time a group of senior Israeli foreign ministry officials took the lead in engaging with the ENP, and former foreign minister Tzipi Livni soon became the champion of the cause.[6]

So far, results have been impressive. While both sides are discussing cooperation on almost every possible economic issue, they signed or initialled several important agreements in recent years. Thus, Israel joined the European satellite navigation programme Galileo in 2004. In April 2008, both sides initialled a preliminary agreement on further trade liberalization in agriculture and fishery, a field in which the EU has traditionally been rather protectionist,[7] while starting negotiations on the liberalization of services. A liberalization agreement pertaining to the aviation sector was signed in February 2008, laying the foundation for an open-skies policy, and negotiations on a more comprehensive agreement are on-going.[8] In July 2008 Israel joined the EU's academic exchange programmes Tempus and Erasmus Mundus. The country is also part of the EU's Copernicus project, which develops civilian satellites, and it participates in a project of the European Space Agency. Similarly significant, a protocol on the general principles that will govern Israel's participation in EU-internal programmes was adopted in April 2008; Israel thus became the first ENP country to sign such an agreement.

Israel's increasing integration into the EU's internal market entails the approximation of Israeli norms and standards to those of the EU in a growing number of fields, a process that Brussels also supports financially (Commission, 2009: 17).[9] This development is advantageous to Israel since European or third-country companies may increasingly benefit from the EU's single market by locating themselves *in Israel*. This is particularly significant in fields in which Israel has a comparative advantage, such as high tech, phytosanitary products, or food-processing (Tovias and Magen, 2005: 421). Of course, especially considering Israel's advanced high-tech sector, together with the EU's traditional trade surplus with Israel, deeper economic relations are in the Union's interest as well.

In March 2007, Israel requested an additional upgrading of bilateral relations, to which Brussels responded favourably during the June 2008 Association Council. Requesting an 'advanced status' in the ENP framework, Israel demanded to take part in a large number of EU policies and programmes. While the government in Jerusalem had also suggested participating in the EU's Council meetings – much to Brussels' surprise – the upgrade envisaged, inter alia, ad hoc summits between Israeli and European prime ministers, and three meetings a year at foreign minister level (*Jerusalem Post*, 8 December 2008). For the then foreign minister Livni, this development reflected the strengthening cooperation between Israel and the EU, 'based on common values and similar world views' (*Jerusalem Post*, 9 December 2008).

Following the Gaza war of winter 2008–09 the upgrade was put on hold. Yet, the freeze, on which there is no official document, only concerns the upgrading of relations agreed upon in June 2008. Whatever was agreed beforehand is not affected. Thus, the agreement on agriculture mentioned above was signed in November 2009 (*Jerusalem Post*, 6 November 2009). Concurrently, the European commissioner for transport Antonio Tajani visited Israel shortly after the end of the Gaza war to discuss future cooperation on transportation and space (Ministry of Foreign Affairs of Israel, 2009). And the annual meetings of the subcommittees established under the ENP resumed regularly in the winter of 2009, after the anticipated hiatus of, indeed, one year. Thus, throughout 2009 the EU and Israel were again playing the 'double game of economic passion and political hostility' (Pardo, 2004).

## Divergence between Rhetoric and Practice

Political divergences are easily picked up by the media in both Israel and Europe, leading to a partially distorted picture of EU–Israeli relations. Particularly during the first years of the second *intifada*, the EU's repeated calls to stop the excessive use of force, refrain from extra-judicial killings, facilitate humanitarian assistance to the Palestinians, and reverse the settlement policy (Council, 2002: 14–15; 2003: 34–5) usually received sharp official replies from Jerusalem (Dachs and Peters, 2005).

Interestingly, the qualitative improvement of EU–Israeli economic ties of recent years not only occurred amid protests from the Palestinian Authority and Arab states; representatives of the EU and its member states repeatedly voiced their criticism of Israeli policies as well. These incidents were widely reported, such as the EU's public condemnation of Israel's continuous settlement expansion under the government of Ehud Olmert (AFP, 14 March 2008). Another example was Brussels' denouncing of Israel's 'disproportionate use of force' (AFP, 2 March 2008) after the Israeli army killed 66 people in Hamas-ruled Gaza in response to rocket firing into southern Israel in March 2008.

On the Israeli side, an editorial in the *Jerusalem Post* of June 2008 summed up persisting Israeli mainstream perceptions of 'the Europeans'. Praising the envisaged upgrade of relations because of the expected economic benefits for Israel, the editorial stressed that it would complement the EU's financial contribution to the Palestinian Authority while countering the EU's 'problematic engagement' with the region.[10] According to the editorial, the upgrade also promised a stronger Israeli leverage over the EU 'to take a firm stand against Hamas and Iran while coaxing Palestinian moderates to temper their demands so as to increase chances of a bargaining breakthrough' (*Jerusalem Post*, 19 June 2008).

With regard to the Gaza war, the then prime minister Olmert had initially praised European leaders for their 'extraordinary support for the state of Israel' (quoted in Gresh, 2009), but the public exchange of accusations returned once the extent of the devastation in Gaza became visible. On Israel's refusal to lift the Gaza blockade, former EU external relations commissioner Benita Ferrero-Waldner (2009) stressed in an article published in the Israeli daily *Haaretz* that 'holding a population of 1.5 million Palestinians hostage for acts, however dangerous and illegal, over which they have no control' was 'unhelpful'. She also regretted that EU subsistence allowances to Gaza were blocked because the Israeli government did 'not allow sufficient cash to enter Gaza'. In return, Israel's Foreign Ministry asked the EU to 'keep a low profile and conduct a quiet dialogue', threatening along the lines of Aba Eban's 1973 statement that 'if these declarations continue, Europe will not be able to have involvement in the peace process' (AFP, 30 April 2009). The EU's initial endorsement of the contentious Goldstone report, which condemned Israel's military offensive in Gaza and accused both Israel and Hamas of deliberate war crimes, also triggered an exchange of public EU–Israeli accusations.[11] Similarly, the media extensively reported on the harsh reactions in Brussels and European capitals to Netanyahu's initial refusal to endorse the two-state solution. Ferrero-Waldner (2009) remarked that 'undermining the viability of a negotiated settlement, in particular by expanding illegal settlements and security perimeters, is unhelpful' ('unhelpful' apparently being Brussels' ultimate expression of disapproval).

The suggestion of former EU foreign policy chief Javier Solana of July 2008 to set a deadline for recognizing the State of Palestine if negotiations failed to reach an agreement, and Israeli foreign minister Lieberman's public bashing of this idea, also attracted much attention. The media also reported extensively on Israel's furious reaction to the comment of an official from the EU representative office in East Jerusalem stating that European taxpayers paid most of the price of Israel's settlement policy, which contributed to the strangling of the Palestinian economy (Keinon, 2009). However, while mutual accusations have become more frequent, and far more acid, since Netanyahu took office, the rhetoric still stands in contrast to the practice of EU–Israeli relations, as we have seen.

*Diverging Visions of Peacemaking*

As the discussion so far indicates, the most vociferous disputes in EU–Israeli relations revolve around the issue of peacemaking, which constitutes the third pattern. These disagreements are particularly salient whenever a right-wing government is in power in Jerusalem. EU–Israeli differences on these matters may well be anchored in diverging *Weltanschauungen* (Dror and Pardo, 2006: 35ff.), resulting from different histories and experiences. Yet while heads of governments obviously also change on the European side, with some being more sympathetic to Israeli positions than others, collective stances on Middle East peacemaking have, in fact, not changed greatly since the EC issued its Venice Declaration. Thus, Brussels has always expressed its commitment to Israel's security and right to exist, but it disputes the legality of Israel's control over the territories acquired in 1967, and it considers Israeli settlements in the occupied territories as illegal under international law. This includes East Jerusalem, as Brussels reiterated in a recent Council Conclusion (2009). Moreover, while being committed to the two-state solution, the EU also considers the separation barrier, where built on occupied land, together with house demolitions and evictions, as illegal under international law, in addition to being an obstacle to peace.

It is important to note that the EU and its member states have invested greatly in this vision for peacemaking ever since the Oslo process started. Back then, the Europeans were in accordance with the Israeli government, and the EU and its member states subsequently became the largest international donors to the Palestinian Authority in support of infrastructure and institution-building. With the outbreak of the second *intifada*, which also entailed the destruction of EU-funded infrastructure in the Palestinian territories, EU assistance shifted to humanitarian aid. The boycott of the Hamas-elected government in 2006 did not stop Brussels from channelling funds to the Palestinians in Gaza while bypassing Hamas, and humanitarian aid actually increased.[12] Moreover, the EU has increased its security role in the region. Thus, as noted above, the EU provided third-party supervision at the Rafah crossing after Israel's withdrawal from Gaza (the mission is suspended since Hamas took over Gaza in 2007). In the West Bank, the EU has been training Palestinian police officers, criminal prosecutors, and judges in the framework of its

EUPOL-COPPS mission. And after the Israel–Hizballah war of August 2006, European troops make up the bulk of the revamped UNIFIL force in southern Lebanon. EU officials and European governments have also repeatedly stated that 'Europe' would be willing to contribute troops to an international peacekeeping force in the event of an Israeli–Palestinian peace deal (Eldar, 2009).

Concurrently, European public opinion has grown increasingly critical of Israeli policies toward its neighbours, and Israel's commitment to genuine peacemaking has been questioned.[13] Similarly, European academics and public figures have criticized the logic of 'financing the Israeli occupation' with European money, thus releasing Israel from its responsibilities as an occupying power under international law (Le More, 2008; Cronin, 2009).

In this light, it should come as no surprise that Israeli governments that do not even pretend to engage in peacemaking according to at least some of the EU's criteria – which correspond to international law principles – will find themselves on a collision course with Brussels (and with some European governments as well). Even after Netanyahu expressed his support for an independent Palestinian state at his Bar-Ilan University speech in June 2009, the government's continuous undermining of this principle by the expansion of settlements in the West Bank and in East Jerusalem cannot but strain EU–Israeli relations.[14] The suspension of the upgrading of EU–Israeli relations, along with the current contemplation of some European countries to present Hamas with different conditions for ending the boycott – namely the acceptance of the 2002 Saudi peace initiative – is a case in point. Finally, the events around Israel's raid on the Gaza flotilla in May 2010 have strained EU–Israeli political ties further. On the European side positions such as that of former Swedish minister Pierre Schori (2010) that 'it's time the EU told Israel that enough is enough' were frequent. Denouncing those accusations, the government in Jerusalem – and many Israelis as well – stressed that 'the Europeans' failed to understand that at least some of the alleged 'peace activists' on the flotilla had no peaceful intentions at all, but were Hamas supporters aiming at provoking Israel.

*Bilateral Relations versus Peacemaking Conditionality*

The fourth recurring pattern in EU–Israeli relations is the disagreement over whether bilateral ties should be conducted in isolation from regional politics, as Israeli governments prefer, or whether they should be linked to the achievement of an Israeli–Palestinian peace settlement.

Brussels sought to link economic concessions toward Israel to progress in the realm of peacemaking, or at least to the display of Israeli good intentions (since it obviously also takes the Palestinian side to reach a deal). Indeed, as noted above, until the advent of the Oslo process Brussels refused to advance economic relations because of Israel's settlement policy. The EU's subsequent affirmation of Israel's 'special status' in its 1994 Essen declaration – much to Israel's satisfaction – gave way to negotiations on a new trade agreement, leading to the 1995 Association Agreement. Implying a linkage to regional politics, the EU's *a posteriori* linking of this agreement to the EMP that had just been launched caused some discontent on the Israeli side. EU–Israeli dissonance on this specific issue was a sign of things to come.

Brussels also considered the EMP a means to increase its political role in Middle East peacemaking, whereas Israel sought to prevent exactly that. While Israel preferred to keep the US as the main broker – the US had proved to be much more favourable to Israeli positions than Brussels – governments in Jerusalem opposed the idea of making the advancement of bilateral relations to the EU conditional on progress in the peace process (Del Sarto and Tovias, 2001). Giving in to the Israeli position, the EMP remained *formally* separated from the peace process. De facto, however, progress in the latter determined the progress of the EMP's regional track – along with the extent of EU–Israeli harmony, as we have seen.

However, Israeli governments also saw the advantages of enhanced regional cooperation under the EMP. Defying the position of Arab states that the normalization of relations with Israel depended on the reaching of an encompassing peace agreement, Israel considered its participation in regional projects as a diplomatic success since it seemed to increase the country's legitimacy. Israel also insisted on its 'special' status vis-à-vis Brussels, constantly lobbying for a separation of bilateral ties from regional politics. While Israel's economic, technological and political features cannot be compared to those of its neighbours, this position characterizes all Israeli governments, irrespective of their political colour. Unsurprisingly, Israel was very pleased when Brussels launched the ENP in 2003–04. The policy not only promised to cut the linkage between bilateral ties and peacemaking that Israel so detested. It also offered the country the opportunity to take full advantage of the EU's offer of a 'stake in the EU's internal market' precisely because of its advanced status and well-developed relations to Brussels – as indeed happened.

The EU, on the other hand, has been utterly incoherent on the question of how to strike a balance between bilateral ties and regional policies in its southern periphery. And nor has it been able to reconcile its political ambitions with its economic and security interests, or its allegedly normative principles with its multiple historical legacies. The ENP, which was originally conceived for the eastern 'neighbours' and ended up in the southern Mediterranean more by accident than by design (Del Sarto and Schumacher, 2005) is a case in point. While this policy maintains an uneasy relationship with what can be termed the region-building logic of the EMP (Adler and Crawford, 2006; Del Sarto, 2006), Brussels did not mean to completely cut off bilateral ties from Middle East peacemaking. In this vein, the ENP's EU–Israel Action Plan of 2004 refers to peacemaking on the basis of the two-state solution and the respect for international law (Commission, 2004) – although the wording is extremely ambiguous (Del Sarto, 2007: 61–3). Similarly, the document summing up the meeting of the EU–Israeli Association Council of June 2008, which stipulated the further upgrading of bilateral ties, defines 'the resolution of the Israeli–Palestinian conflict through the implementation of the two-state solution' (European Parliament, 2008: 2) as a common interest and objective. Certainly, there is no consensus among EU member states on linking the future of EU–Israeli economic ties to the fate of peace negotiations. A number of EU member states – most notably Britain, Ireland, Cyprus and Malta – had already unsuccessfully sought to establish this linkage at the end of 2008 (Keinon, 2008). However, behind closed door the EU did formulate conditions for 'unfreezing' the upgrade of relations.[15] These include the commitment to the two-state solution, 'visible evidence of the new government's seriousness in pursuing the path of peace', the lifting of the Gaza blockade and the halt of settlement expansion, house demolitions and evictions in the territories, including East Jerusalem (Ferrero-Waldner, 2009).

The Israeli ambassador to the EU, Ran Curiel, may well downplay the importance of the EU's impatience, claiming that the 'positions of Israel and the EU are actually converging'. However, his assessment that the upgrade is not a 'gift to Israel, but rather of mutual interest to both Israel and the EU' (quoted in Philips, 2009) may not be fully shared in Brussels, as the continuation of the freeze demonstrates. In fact, it seems increasingly clear that talks with the Palestinians on a two-state solution afford Israel its sole defence against eroding relations with Europe in the long run, as former Israeli ambassador to the EU Oded Eran (2009) recently noted.

**The EU and Israel: Patterns, Positions and Ambiguities**

One explanation for EU ambivalence towards Israel is that economics drive policy making in Brussels (Miller, 2006: 657–9). Certainly, much of Israeli industrial production is in high added-value sectors, and EU countries benefit from Israel's advanced technologies. Israel's participation in the EU's research and development programmes must be seen in this light. But this is only one part of the explanation. In fact, economic policies are relatively uncontested within the EU, and a wide array of proven instruments and Community-driven policies are in place. Moreover, Israel is geographically close to Europe, and its economy is comparable to those of EU member states (Israel has just been admitted to the OECD). The country also has a well-developed infrastructure and business environment, while maintaining a wide array of cultural relations with most European countries. All these factors are conducive to the smooth development of economic ties, which do not tend to make headlines – much in contrast to the disputes over the Arab–Israeli conflict.

The EU's internal consensus on economics contrasts with the positions of the various member states on Middle East peacemaking. Although, as noted above, the EU's stance on this issue has not changed considerably over the years, it still remains the lowest common denominator among the member states, with some members being far more critical of Israeli policies than others.[16] To complicate matters further, the EU is also striving to play a greater role in world politics, with the Middle East as the EU's alleged 'backyard' being a welcome arena. But while the Union's instruments and institutional set-up hardly match those ambitions, these conflicting aims result in inconsistent policies toward the Middle East and North Africa, such as the (regionally oriented) EMP/UfM and the (bilaterally oriented) ENP.

Israel's strategy, on the other hand, has been pretty linear. It does not trust 'the Europeans' when it comes to peacemaking and Israel's security, and it is convinced that the history of relations has repeatedly proved it right. Moreover, according to Israeli officials, the EU's loud-voiced declarations on what Israel should and should not do hurts its credibility, along with its standing in Israeli public opinion. Furthermore, governments in Jerusalem clearly perceive the EU to be adopting double standards in its Mediterranean policy, such as repeatedly criticizing Israel, but rewarding Morocco with an 'advanced status' in spite of the lack of democracy, or pampering Tunisia notwithstanding its grim human rights records. Israel's main preoccupation has thus been to develop its bilateral economic relations to the EU. The country is also ready to concede Brussels *some* political involvement in Middle East peacemaking, but only provided that the latter shows sensibility to Israel's security concerns – and stops its criticism.

## Israel and the Union for the Mediterranean

Considering the patterns of EU–Israeli relations discussed so far, what was Israel's position and strategy towards the emerging Union for the Mediterranean? Given that Israel had profited considerably from the decoupling of bilateral ties from regional politics under the ENP, the reanimation of the regional Barcelona Process in the form of the UfM should have received a rather lukewarm Israeli response at best. In fact, it could also have been expected that Israel would decide not to participate in the UfM at all, insisting on the bilateral framework for its relations to the EU instead. Reflecting the traditional Israeli position toward, and experience with, the regional track of the EMP, in private Israeli officials were indeed rather sceptical. Yet in public, the Israeli government under Olmert praised the initiative and pledged the country's active participation. The main reason was that the UfM was the pet of Sarkozy, who, in reversal of French policies, had been particularly supportive of Israel (Pardo and Peters, 2010: 44). Indeed, Paris had played an exceptionally important role in sustaining Israel's request for the upgrade in EU–Israeli relations, and Brussels' favourable response of June 2008 occurred, indeed, under the French EU Presidency.

The development of bilateral ties of recent years also permitted Israel to consider the advantages of the UfM with a greater detachment. From an Israeli perspective, Euro-Mediterranean regional cooperation no longer seemed to contradict Israel's 'special status' vis-à-vis Brussels. Secondly, according to Israeli officials, Israel did not want to be the first country turning down an EU initiative. Thirdly, the Israeli government may also have hoped that all sides (and particularly the EU) had learned from the EMP experience. Specifically the project-based nature of the UfM seemed to support such expectations, as it seemed to entail a lower degree of politicization. At the same time, Israel considered that it could make an important contribution to regional cooperation under the UfM, as it has considerable experience in the priority fields of the envisaged projects, such as solar energy and the environment. These aspirations reflect the country's traditional position that, independently of Arab–Israeli peacemaking, regional cooperation fostering stability and development is in Israel's interest as well. In this vein, Israeli diplomats stressed that the UfM meeting on the status of women, which took place in Morocco in November 2009, was important per se, and unrelated to peacemaking. Finally, as with the EMP beforehand, the government in Jerusalem reckoned that the envisaged cooperation with Arab states under the UfM could enhance Israel's legitimacy among them, even before a formal peace agreement was reached. Thus, for Israel it was a success that Arab states that have not recognized the country (such as Syria) continued to sit with it in the same room in the UfM framework.

The Israeli government hence decided to go along with the initiative, which it did not want to boycott or undermine for strategic reasons. Particularly as regards Sarkozy, but also in view of the excellent development of its economic ties with the EU, Israel thus played the role of a 'favour exchanger' (see Bicchi, this collection). Yet the country also sought to influence the evolving initiative according to its preferences, while seeking to veto what it perceived as running against its interests. Most noticeably, from the outset Israel adamantly opposed the involvement of the Arab League, out of concern that the UfM would become more politicized, with Israel finding itself in an isolated position. Indeed, in the past, Arab states had occasionally prevented Israel's participation in EMP meetings out of protest against Israeli policies (see Bicchi, this collection). Reflecting yet another long-standing position explained above, Israel also wanted to avoid the Arab–Israeli conflict becoming a legitimate concern of a Brussels-led regional initiative, possibly with a growing political role for the EU. The Israeli government subsequently agreed to an observer status for the Arab League, but insisted that the latter would not have the right to participate at all meetings and to speak at all levels. In exchange for the post of one of the six deputy secretary-generals of the UfM, which was considered a diplomatic achievement in Jerusalem, Israel eventually gave up its resistance. Thus, it was agreed eventually that the Arab League would participate at all UfM meetings, with the right to speak but not to vote. The dispute over the Arab League's involvement, together with the mainly Euro-Arab quarrel over where to base the UfM secretariat,[17] caused a deadlock lasting several months.

Of course, as with the idea of a 'New Middle East' during the Oslo years, it was always unlikely that the majority of Arab states would accept Israel's potential contribution in the envisaged projects, which, as in the past, may raise fears of Israeli regional hegemony. Moreover, the Arab states' position on normalization in the absence of a comprehensive peace agreement has remained unaltered. This stance is even more unlikely to change given that Israeli governments have so far plainly ignored the 2002 Saudi peace initiative that offers full peace in exchange for an Israeli withdrawal to the 4 June 1967 borders. The Arab League, which repeatedly endorsed the Saudi peace proposal, it should be stressed, is now formally part of the UfM. And perhaps unsurprisingly, the organization promptly stressed that its participation in the UfM would not imply the normalization of relations with Israel (AFP, 4 November 2008).

Following the Gaza War, with the UfM at a virtual standstill, disappointment set in on the Israeli side. The greater say of the 'southern' states that the concept of co-ownership has a price, with the Egyptian co-presidency delaying the constitution of the UfM's governing bodies, 'starting with the Secretariat in protest against Israel's invasion of Gaza at the end of 2008' (Aliboni, 2009: 3). The Secretariat was only nominated in January 2010, but concrete decisions on UfM projects continued to be held up. Some meetings at the ministerial level took place in the meantime, but with unsatisfactory results. For instance, Arab–Israeli disputes prevented the adoption of a regional water strategy during the fourth Euro-Mediterranean Ministerial Conference on Water, which took place in April 2010 in Barcelona. In this case, there was agreement on all aspects of the document, except for the term 'occupied territories', which Israel opposed and on which the Arab League insisted – although Israel proposed 'territories under occupation' as a compromise (*Jerusalem Post*, 16 April 2010). The meeting of foreign ministers scheduled to take place in Istanbul in November 2009 was cancelled, as the Arab participants refused to meet Israel's foreign minister Avigdor Lieberman. Lieberman, it should be noted, not only embraces extreme positions on Israel's Arab citizens, but had also said in the past that Egyptian President Hosni Mubarak could 'go to hell' (and had also suggested bombing Egypt's Aswan Dam).

Lieberman's convictions and personality may constitute a specific problem for Euro-Mediterranean relations. A settler himself, he believes that the Oslo process signalled the decline of Israel's international status, as he stated upon assuming the position of foreign minister (Eldar, 2009). While publicly proclaiming 'an exchange of populated territory' as part of a Middle East peace deal (*Haaretz*, 29 September 2010),[18] he also quite undiplomatically told French and Spanish foreign ministers Bernard Kouchner and Miguel Angel Moratinos at a dinner in Jerusalem to 'solve their own problems before they complain to Israel' (*Haaretz*, 11 October 2010). Yet, as we have seen, disagreements over the modalities of peacemaking are a central feature of EU–Israeli relations, and all Israeli governments dislike European involvement in the peacemaking arena – with the dislike of a right-wing government tending to be greater. But the main problem is that the UfM's institutional set-up is conducive to any type of political dispute disrupting its proceedings. Together with the involvement of the Arab League, the decision to have both an Israeli and a Palestinian among the six UfM deputy secretary-generals is actually tantamount to begging for trouble. This arrangement is likely to additionally block, or in the best case delay, the implementation of UfM projects – at least as long as the Arab–Israeli conflict is not solved. Hence, in the current political climate, the UfM will continue to be held hostage to the Arab–Israeli conflict.

## Conclusions: *Plus ça change* ...

The patterns characterizing EU–Israeli relations discussed above shed light on Israel's position and strategy towards the UfM while assessing the relevance of the latter for EU–Israeli relations. The most important fact is that the change in EU–Israeli relations over recent years resulted from the impressive advancement of *bilateral* ties under the ENP. This development also entailed a temporary softening of the tone of EU–Israeli disputes, along with a growing Israeli acceptance of a (moderate) European security role in the region. Although political relations are currently under stress, governments in Jerusalem continue to consider their bilateral ties to Brussels as the mainstay of EU–Israeli relations, as in the past. Regional cooperation under the UfM (or in any other form) is only an optional side-show for Israel. In view of this very clear set of priorities, Israel's participation in the emerging UfM followed mainly strategic considerations. It first acted as a 'fence-sitter' and 'favour exchanger' as regards specifically Sarkozy, but also vis-à-vis the EU as a whole. Israel subsequently sought to fight the UfM's transformation into a legitimate forum for dealing with Arab–Israeli peacemaking under EU auspices, much in accordance with its traditional position.

Secondly, the development of EU–Israeli bilateral relations of recent years clearly demonstrates that Brussels only very timidly and inconsistently, if at all, embraces the use of economic concessions toward Israel as potential leverage in the peacemaking arena. In fact, EU–Israeli disagreements over Middle East peacemaking accompanied the qualitative improvement of bilateral economic relations on the ground. The aftermath of the Gaza war, which witnessed harsh European criticism at Israel's violation of international law, but, with the exception of the 'mutually agreed' freeze of the upgrade, also 'business as usual', most clearly proves this point. True, the planned upgrade of EU–Israeli bilateral relations is unlikely to occur in the current political climate. But it is also improbable that the agreements signed so far, and the growing institutionalization of relations, will be cancelled or reversed. In addition to a basic European schizophrenia vis-à-vis Israel, which is anchored in recent history, the main reason here is that the EU is internally divided, which effectively prevents the necessary decision by consensus among EU member states on any form of conditionality, let alone sanctions. This fact is far more relevant than any current (and future) EU criticism of Israeli policies, whether in the UfM framework or beyond.

Thirdly, the launching of the UfM entailed an attempt to rebalance the EU's Mediterranean policy from its bilateral excesses under the ENP back toward the regional dimension. The reality is, however, that bilateralism has effectively overwritten the regional logic in the EU's Mediterranean policy, as the development of EU–Israeli relations demonstrates. Moreover, it is questionable whether the revamped Barcelona process/UfM will provide the appropriate forum for the EU's political ambitions. The short history of the UfM has shown that its institutional set-up has in fact increased the potential for politicization as regards the Arab–Israeli conflict, and thus for deadlock. At the same time, the deepening of relations with Israel without applying any sort of conditionality (unlike, say, in the case of Hamas), erodes the EU's legitimacy among Arab states and the Palestinians. Again, this is far more pertinent than any EU concession to the Arab side, such as the Arab League's participation in the UfM.

Finally, disagreements over the modalities of peacemaking will not only determine the future of the UfM, but will also continue to be a major issue in EU–Israeli relations. In the current situation, the UfM is likely to become a prominent forum for those disputes, potentially strengthening the Israeli perception that the UfM as just another forum for 'Israel-bashing' and thus possibly hardening Israel's traditional opposition toward a greater EU role in Middle East peacemaking. Given that Brussels has invested extensively in this realm, prolonged disputes with Israel over this issue are likely to affect bilateral relations in the long run, particularly if there is no change of government (or policies) in Jerusalem. In the short and medium term, however, the EU is internally too divided to impose conditions on Israel – notwithstanding occasional threats to the contrary from individual European politicians. Bilateral cooperation will advance, with the media continuing to ignore this side of EU–Israeli relations while extensively covering the mutual accusations. Thus, while Israel's position toward the emerging UfM, and the role it has been playing so far, was hardly surprising considering the patterns of EU–Israeli relations, the UfM is unlikely have a major impact on the latter. Hence, it seems that the more things change in the EU's Mediterranean policy, the more they remain the same.

## Acknowledgements

The author would like to thank Federica Bicchi, Richard Gillespie, Annette Jünemann, Emma Murphy, Chiara Steindler, and Alfred Tovias for their insightful comments on earlier drafts of this paper. She would also like to express her gratitude to the Pears Foundation for their continuous support. The usual disclaimers apply.

## Notes

[1] Germany and the European Commission were particularly opposed to Sarkozy's idea of a separate and intergovernmental policy initiative. On the genesis of the UfM see Bechev and Nicolaidis (2008) and Balfour (2009); on Germany's role see Schumacher (this collection).

[2] Israel opened a diplomatic representation to Brussels in April 1958.

[3] Together with substantial funds, the 1999 Berlin Declaration under the German presidency was apparently aimed at 'compensating' the Palestinian Authority for delaying the envisaged unilateral declaration of an independent Palestinian state.

[4] Under considerable international pressure, Israel eased the blockade in June 2010.

[5] Among the ENP partner states in the south, Morocco also pushed for deeper relations with the EU from the outset, much in accordance with its traditional policy. In October 2008 Brussels awarded the country an 'advanced status' (Martín, 2009) and in October 2010 it was also awarded to Jordan.

[6] Ironically, European officials initially perceived the Israeli fervour as a bit too excessive.

[7] Morocco has also been enjoying the liberalization of trade in agriculture and fisheries in its relations with the EU.

[8] See the website of the EC's DG Transport at http://ec.europa.eu/transport/air_portal/international/pillars/common_aviation_area/israel_en.htm (accessed 15 November 2010).

[9] Brussels supported this process with €8 million for the period 2007–10.

[10] The editorial specifically mentions the EU's backing of Palestinian and Israeli NGOs that 'pursue partisan activities' with 'less than benign influence on the conflict', such as the Committee against House Demolitions and Adalah, the centre for Arab minority rights in Israel.

[11] In the subsequent voting in the UN General Assembly on the non-binding resolution recommending that the Goldstone report be referred to the UN Security Council, Germany, Italy, the Netherlands, Poland, Hungary, the Czech Republic and Slovakia voted against the resolution. Austria, Britain, France, Bulgaria, Denmark, Latvia, Sweden, Romania, Greece, Belgium, Estonia, Latvia, Lithuania, Finland, Luxembourg and Spain abstained. Malta, Ireland, Portugal, Slovenia and Cyprus voted in favour of the resolution.

[12] According to EU figures, between 2000 and 2009 the European Commission sent over €3.3 billion in aid to the Palestinians. In 2007, when Hamas took control over Gaza, the annual aid peaked to €563 million. These figures do not comprise the contributions of single EU member states. See http://ec.europa.eu/external_relations/occupied_palestinian_territory/ec_assistance/eu_support_pa_2000_2009_en.pdf (accessed 18 June 2010).

[13] It would be important to study public opinion on Israel and the Israeli–Palestinian conflict in EU countries, and the possible impact on policy decisions. However, this complex issue cannot be treated here.

[14] Under considerable pressure from Washington, Netanyahu agreed on a 10-month settlement freeze in November 2009, which, however, did not include East Jerusalem. Israeli newspapers reported that construction in the settlements in the West Bank was continuing despite the freeze (Eldar, 2010). So far, the freeze has not been renewed.

[15] This move was contested by some EU members. Czech premier Topolek openly challenged Ferrero-Waldner's attempt to introduce conditionality in EU–Israeli relations during the EU Czech Presidency (*Haaretz*, 26 April 2009).

[16] The voting of the single EU member states on the Goldstone Report mentioned earlier is a case in point.

[17] Ministers from the southern rim eventually agreed to back the Spanish city of Barcelona as the seat of the UfM secretariat in exchange for the post of secretary-general going to a 'southern' member.

[18] Before the UN General Assembly in September 2010, Lieberman suggested ceding parts of Israel with large Arab populations (Israeli citizens) to a future Palestinian state in exchange for Israel keeping large settlement blocs in the West Bank, a proposal which has been part of his party's platform.

## References

Aliboni, R. (2009) *The Union for the Mediterranean: Evolution and Prospects* (Rome: Istituto Affari Internazionali, Documenti IAI 0939E).

Adler, E. & Crawford, B. (2006) 'Normative power' and the European practice of region building: the case of the Euro-Mediterranean Partnership, in: E. Adler, F. Bicchi, B. Crawford & R. A. Del Sarto (Eds) *The Convergence of Civilizations: Constructing a Mediterranean Region*, pp. 3–47 (Toronto: University of Toronto Press).

Balfour, R. (2009) The transformation of the Union of the Mediterranean, *Mediterranean Politics*, 14(1), pp. 99–105.

Bechev, D. & Nicolaidis, K. (2008) The Union for the Mediterranean: a genuine breakthrough or more of the same? *The International Spectator*, 43(3), pp. 13–20.

Commission of the European Communities (2004) EU/Israel Action Plan, Final (published 9 December 2004). Available at http://ec.europa.eu/world/enp/pdf/action_plans/israel_enp_ap_final_en.pdf (accessed 3 December 2009).

Commission of the European Communities (2009) Progress report Israel, Brussels, 23 April, SEC(2009) 516/2. Available at http://ec.europa.eu/world/enp/pdf/progress2009/sec09_516_en.pdf (accessed 24 January 2011).

Council of the European Union (1994) Extracts of the conclusions of the Presidency of the Essen European Council, 9 and 10 December, Bulletin of the European Union, Supplement 2/95, p. 28.

Council of the European Union (2002) European Council declaration on the Middle East, Copenhagen European Council, 12–13 December, Presidency conclusions, Annex III.

Council of the European Union (2003) Brussels European Council 20 and 21 March 2003, Presidency Conclusions.

Council of the European Union (2009) Council conclusions on the Middle East Peace Process, 2985th Foreign Affairs Council meeting, Brussels, 8 December. Available at http://www.consilium.europa.eu/ueDocs/cms_Data/docs/pressData/en/gena/105545.pdf (accessed 24 January 2011).

Council of the European Union (2010a) Joint Statement by high representative Catherine Ashton and minister for foreign affairs of the Russian Federation Sergey Lavrov, Brussels, 1 June 2010, A 96/10. Available at http://www.consilium.europa.eu/uedocs/cms_Data/docs/pressdata/EN/foraff/114733.pdf (accessed 24 January 2011).

Council of the European Union (2010b) Council conclusions on Gaza, 3023rd Foreign Affairs Council Meeting, Luxembourg, 14 June 2010. Available at http://www.consilium.europa.eu/uedocs/cms_Data/docs/pressdata/EN/foraff/115158.pdf (accessed 18 June 2010).

Cronin, D. (2009) Q & A: EU stepping closer to Israel, regardless, Interview with Benita Ferrero-Waldner, *IPS Inter Press Service*, 29 July.

Dachs, G. & Peters, J. (2005) Israel and Europe, the troubled relationship: between perception and reality (Beer Sheva: Ben-Gurion University, the Centre for the Study of European Politics and Society), in:

R. Nathanson & S. Stetter (Eds) *Reader of the Israeli-European Policy Network* (Tel Aviv: Friedrich Ebert Stiftung), pp. 317–33.

Del Sarto, R. A. (2006) *Contested State Identities and Regional Security in the Euro-Mediterranean Area* (New York and Basingstoke: Palgrave Macmillan).

Del Sarto, R. A. (2007) Wording and meaning(s): EU–Israeli political cooperation according to the ENP Action Plan, *Mediterranean Politics*, 11(1), pp. 59–74.

Del Sarto, R. A. & Schumacher, T. (2005) From EMP to ENP: what's at stake with the European Neighbourhood Policy towards the Southern Mediterranean?, *European Foreign Affairs Review*, 10(1), pp. 17–38.

Del Sarto, R. A. & Tovias, A. (2001) Caught between Europe and the Orient: Israel and the EMP, *The International Spectator*, 36(4), pp. 61–75.

Dror, Y. & Pardo, S. (2006) Approaches and principles for an Israeli grand strategy towards the European Union, *European Foreign Affairs Review*, 11(1), pp. 17–44.

Eldar, A. (2009) Continental divide, *Haaretz*, 23 April.

Eldar, A. (2010) Only an idiot would say Israel has frozen settlement activity, *Haaretz*, 26 January.

Eran, O. (2009) A reversal in Israel–EU relations?, *Strategic Assessment*, 12(1) (Tel Aviv: Institute for Strategic Studies). Available at http://www.inss.org.il/upload/(FILE)1244445178.pdf (accessed 1 December 2009).

European Parliament (2008) Eight Meeting of the EU-Israel Association Council, Luxembourg, 16 June, Statement of the European Union. Available at http://www.europarl.europa.eu/meetdocs/2004_2009/documents/dv/association_counc/association_council.pdf (accessed 24 January 2011).

Ferrero-Waldner, B. (2009) The offer on the table, *Haaretz*, 17 April.

Gresh, A. (2009) Gaza war changes Middle East equation at Israel's expense, *Le Monde Diplomatique*, online edition, February. Available at http://mondediplo.com/2009/02/02gazawar (accessed 26 October 2010).

Keinon, H. (2008) Amid displeasure with UK, Olmert to go to London, *Jerusalem Post*, 10 December.

Keinon, H. (2009) Israel fumes at EU for saying its taxpayers 'bear settlement burden', *Jerusalem Post*, 7 July.

Le More, A. (2008) *International Assistance to the Palestinians after Oslo: Political Guilt, Wasted Money* (London: Routledge).

Martín, I. (2009) EU–Morocco relations: how advanced is the 'advanced status'?, *Mediterranean Politics*, 14(2), pp. 239–245.

Miller, R. (2006) Troubled neighbours: the EU and Israel, *Israel Affairs*, 12(4), pp. 642–664.

Ministry of Foreign Affairs of the State of Israel (2000) Reactions by prime minister Netanyahu and foreign minister Sharon on the EU statement on Jerusalem, 25 March 1999, in *Israel's Foreign Relations: Selected Documents*, 17: 1998–1999, Document 154 (Jerusalem: Ministry of Foreign Affairs).

Ministry of Foreign Affairs of the State of Israel (2009) Vice-president of EU Commission visits Israel, 27 January. Available at http://www.mfa.gov.il/MFA/About+the+Ministry/MFA+Spokesman/2009/Press+releases/EU_commissioner_for_transport_visits_Israel_27_Jan_2009' – and please add ' (last accessed 24 January 2011).

Pardo, S. (2004) Narrowing gaps, *The Jerusalem Post*, 2 January, p. 9.

Pardo, S. & Peters, J. (2010) *Uneasy Neighbours: Israel and the European Union* (Lanham, MD: Lexington Books).

Philips, L. (2009) Israel believes war won't harm EU relations upgrade, *EU Observer*, 7 January. Available at http://euobserver.com/9/27357 (accessed 3 March 2010).

Sachar, H. M. (1999) *Israel and Europe: An Appraisal in History* (New York: Alfred A. Knopf).

Shori, P. (2010) It's time the EU told Israel that enough is enough, *Europe's World*, 15 (Summer), pp. 128–33.

Schumacher, T. (2009) A fading Mediterranean dream, *European Voice*, 16 July.

Tocci, N. (2009) *Active but Acquiescent: The EU's Response to Israel's Military Offensive in the Gaza Strip* (Copenhagen: Euro-Mediterranean Human Rights Network, May).

Tovias, A. (2006) Exploring the 'pros' and 'cons' of Switzerland's and Norway's model of relations with the EU: what can be learned from these two countries' experience by Israel?, *Cooperation and Conflict*, 41(2), pp. 203–222.

Tovias, A. (2007) Spontaneous vs. legal approximation: the Europeanization of Israel, *European Journal of Law Reform*, IX(3), pp. 485–501.

Tovias, A. & Magen, A. (2005) Reflections from the new near outside: an Israeli perspective on the economic and legal impact of EU enlargement, *European Foreign Affairs Review*, 10(5), pp. 399–425.

# The Ties that do not Bind: The Union for the Mediterranean and the Future of Euro-Arab Relations

OLIVER SCHLUMBERGER
Institute of Political Science, Eberhard-Karls Universität Tübingen, Germany

ABSTRACT *What impact does the Union for the Mediterranean (UfM) have on the future evolution of Euro-Arab relations? This contribution first reflects on Arab reactions to the UfM, and subsequently analyses what alterations the UfM brings to existing Euro-Arab relations in terms of actors, institutional arrangements, and policy contents. In sum, the UfM caters well to Arab regimes' priorities, namely the maintenance of authoritarian rule: The UfM tends to exclude societal voices and leads to a re-governmentalization of relations; the institutional set-up elevates Arab regimes to become formal veto-players, and the prioritized policy areas have – from an Arab regime perspective – the advantage of being de-politicized and stripped of any ambitious macro-political goals such as democratization. The UfM can thus be considered a triple victory for authoritarian Arab rulers in re-shaping their relations with Europe, and casts serious doubts on the hypothesis of the EU acting as a norm entrepreneur.*

Euro-Arab relations form an important part of the European Union's overall foreign policies. Over half of the Arab population, including the politically most important Arab states, is included in the Euro-Mediterranean Partnership (EMP) established by the EU in 1995 and which superseded earlier co-operation schemes under the so-called financial protocols. After the re-launching of the EMP as the Union for the Mediterranean (UfM) in 2008, one pertinent question is how the new design of this partnership will affect the future evolution of Euro-Arab relations. I argue here that despite initially hesitant Arab reactions, the shifts in actors involved, in the new institutional framework as well as in policy content, seem almost tailor-made to Arab regimes' political preferences while defying any European claims of aiming at meaningful social and political reform (let alone transformation) that could lead to enhanced regime convergence between the northern and southern shores of the

Mediterranean (as a pre-requisite of deeper integration), and to enhanced broad-based economic and human development prospects in the Arab region.

A first glance at the addressees of this policy initiative, i.e. the Arab partner countries, demonstrates some key features which characterize this group of states. First, they are all developing nations that figure within the group of middle income countries according to the World Bank's classification; they are thus not among the world's poorest states. However, a second remarkable feature is that they all face relative economic stagnation or decline when compared to other developing regions around the world (e.g. Henry and Springborg, 2001: ch. 1). As is also well established, one – if not *the* – key reason for this fact is the political stagnation that hampers social, economic and political development in this region, as is evidenced by the fact that the Arab region has been the world's most unfree region for decades (UNDP/AFESD, 2005; Heydemann, 2007; Schlumberger, 2007, 2008). Authoritarian power maintenance has been the top priority of Arab political leaders, even in the face of economic and financial crises which have repeatedly brought Arab states close to bankruptcy. This entails a range of serious consequences for neighbouring regions such as the EU. Among such consequences are primarily hard and soft security issues (such as terrorism, illegal migration, transnational organized crime, drug trafficking and trafficking of humans), but also a lack of international policy co-ordination in sectoral domains such as environmental, water, energy, or financial policies in which today transnational co-ordination gains ever greater importance. All of these aspects may have negative impacts on the Arab region's neighbours.

There are thus manifest and very tangible social, economic and political reasons why Europe should take a pro-active interest in how its Arab neighbours are governed. If Europe, then, is the 'normative power' Manners (2002) and his followers tend to see in the EU, or the 'norm entrepreneur' that Pace (2007) searches for in Europe's foreign policies, we should expect the Arab world to be *the* prime addressee of European efforts at exporting its core political norms and values (such as respect for human rights, the rule of law, or pluralistic, inclusive, transparent and accountable modes of political decision making), whether for utilitarian or idealistic reasons.

This contribution examines whether or not this is the case under the new framework of the UfM. It does so by outlining Arab reactions to the initiation of the UfM which – despite the 'Joint Declaration' (2008) signed by both the EU members and the southern Mediterranean partners in Paris – at first clearly was a European (or rather: French) endeavour at re-shaping Euro-Mediterranean relations (next section). These sceptical reactions are contextualized and discussed before I turn to the UfM proper in its core dimensions of actors, institutions and policies (section three). The analysis of these three dimensions leads to two alternative scenarios as regards the future development of EU–Arab relations (section four). This, in turn, enables me to conclude on the UfM's impact on future political, social and economic developments within the Arab countries themselves, as well as for those countries' future role within the larger fabric of Euro-Mediterranean relations (conclusions).

## Arab Reactions to the UfM

Generally, Europe is seen as a reputable partner of Arab countries and a desire often heard in the streets of Cairo, Amman or Algiers is that the EU should play a more prominent role in Middle Eastern affairs of global relevance, such as the Israeli–Palestinian conflict, but in other matters as well. In its regional policies towards the Arab countries, the EU has largely been perceived on the southern shores of the Mediterranean as a welcome antidote to US Middle East policies which at times have been perceived as overly aggressive, little sensitive to the region's peculiarities, and certainly not as occupying the position of what 'the Arab side' would expect from an external power in terms of honest brokerage in regional conflicts.

But the Arab countries have also benefited individually from the EMP and the subsequently established umbrella of the European Neighbourhood Policy (ENP; since 2004) which has brought them high levels of support for national goals in development and policy reform issues that are – to a large extent – in the genuine interest of Arab regimes. Despite some EU rhetoric to the contrary that chiefly seems to address European audiences, Arab countries were given enough room for manoeuvre to engage in civil society programmes funded by the EU without having to cede control over the latter. Conveniently, they were able to effectively exclude the risk of a real empowerment of societal actors who might otherwise have turned into contestants for political power at home.

The UfM, then, rather than threatening to erode these benefits for Arab countries, offered the prospect of augmenting and deepening them, including prospects of filling new and prestigious international positions within the planned UfM Secretariat and in a range of other bodies yet to be established, while at the same time the contents that were envisaged to form the core of new UfM programmes and projects seem to be void of much of the impetus towards political reform which both the EMP framework and the ENP implicitly and explicitly carried with them.

Nevertheless, as in much of Europe, French President Sarkozy's announcement of plans to establish the UfM clearly met with less than enthusiastic reactions from the Arab countries: initial responses ranged from cautious scepticism to outright disapproval. Two issues were at the core of this Arab scepticism and criticism:

(a) Deteriorated Arab–Israeli relations and an unwillingness, on the Arab side, to enter into what might become a creeping normalization of relations with Israel, and
(b) fears about a renewal of what in the past was sometimes perceived as European paternalism.

The most important issue concerned Arab–Israeli relations which, at the time of the announcement of the UfM, were at a low point. The Israeli invasion of Lebanon (known as the 'Summer War' of 2006) had destroyed most of the latter country's infrastructure just two years earlier. Given these tense Arab–Israeli relations and with little left of a former 'peace process', Arab states were particularly 'worried that if they join' this would 'imply a normalisation of bilateral relations' with the

Jewish state (Vucheva, 2008). Nevertheless, there had been an initial willingness to co-operate and find joint solutions. In late 2008, Israel agreed that the Arab League attend all UfM meetings, while Israel was promised the post of a deputy secretary-general in return. However, with Israel's bombardment of the Gaza Strip in December 2008, relations deteriorated once again. Arab governments felt they had to make sure not to send any signals on the international stage that could be interpreted as any sort of improvement of these relations, let alone a 'normalization'.[1] In this respect, Arab countries initially tried to act as veto-players – and they still do, as is evidenced by the frequent postponements and cancellations of meetings due to reasons associated with Israel's policies in the Middle East conflict.

A second key reason for Arab hesitancy in embracing the French idea of re-framing Euro-Mediterranean relations through the UfM stemmed from a certain degree of dissatisfaction with the Barcelona Process as it had evolved since 1995, and more specifically with an at least perceived paternalism that had been inherent to the EMP/ENP from Barcelona through the European Neighbourhood Policy Instrument (ENPI). Again, it was Algeria's foreign minister Medelci who found clear words: 'Relations with the EU are unbalanced and decisions belong to those who now have money and know-how' (Vucheva, 2008). If this language appears rather undiplomatic, then Libya's Colonel Qaddafi expressed Arab concerns about European domination even more forcefully, calling the initiative 'an insult' that was 'taking us for fools'. At a 2008 summit between North African states and Syria, he pointed out that 'we do not belong to Brussels. Our Arab League is located in Cairo and the African Union is located in Addis Abbeba. If they want cooperation they have to go through Cairo and Addis Abbeba' (Reuters Africa, 2008, as quoted in Katz, 2008). Whatever the EU may think about the emanation of its norms and values, the fact that this has not translated into trust and ownership for Europe's endeavours in the South is hard to contradict. As Schwarzer (2009) put it: 'No equal partnership has been established between the northern and southern states. The Mediterranean Union is still seen in the southern region primarily as an EU project, if not as a project of French political interests.' In a similar vein, the reputable Swiss daily *Neue Zürcher Zeitung* found that

> it should have attracted Sarkozy's attention that the interest the Mediterranean countries share in the EU's multi-billion offers has flagged over the years. Rather than thinking about ways of throwing fresh millions southwards over the sea, there should be an examination of the reasons for this disinterest. (Winkler, 2008; author's translation from German)

These reactions point to two facts. First, the omnipresence of the Arab–Israeli conflict in Arab foreign affairs, which may impact negatively on external relations even if it is not the object of concrete policies. This issue can hardly be resolved by the EU as long as it entertains positions towards that conflict which appear primarily informed by transatlantic strategic considerations rather than by its own political standpoint, and for the foreseeable future the EU will have to live with this potentially disruptive element in its Mediterranean relations unless it is prepared to disrupt its relations with the US. Second, Arab reactions have made it clear that the frameworks of both the EMP and the ENP have not been perceived by the Arab side as a partnership among equals. They have not resulted in greater levels of Arab trust in European foreign policies toward the Middle East and North Africa truly aiming at progress for the benefit of the partner countries. While the first point is a fact Europe will have to accept as long as its transatlantic priorities are a given, the second demonstrates that reactions have been critical towards the new initiative regardless of its concrete contents: the UfM promised the southern partners much more of a say in the joint design of policies and programmes than the EMP/ENP framework, and thus for the Mediterranean Partners Countries (MPCs) could potentially represent a significant step forward as regards their own weight and voice in Euro-Mediterranean relations.

While Arab reactions to the French announcements and to the UfM itself during its initial phase (2008/09) leave little room for interpretation, the more analytical question to be addressed below is whether the Arab countries were well-advised in their early criticism of the UfM. *Should* the Arab states be as critical towards the UfM as they have been initially? The answer arguably hinges upon what the UfM's impact on Arab countries will look like, which in turn will shape the future of Euro-Arab relations.

**Core Dimensions of the UfM: Actors, Institutions, Policies**

To assess the UfM's possible impact for future Euro-Arab relations, I refer to its three core dimensions as carved out in the introduction by Bicchi. A first focus is on *actors* who are likely to be decisive driving (or blocking) forces; second, changes to the *institutional set-up* of Euro-Med relations are discussed with a view to their consequences for the Arab partners; third comes the question of politicization/ de-politicization of Euro-Mediterranean relations discussed in the introduction. This (sub-)section will be labelled *policies* because it is the policy domain that demonstrates a de-politicization which is often overlooked by EU foreign policy analysts. While to some extent unclear at the time of writing (late 2010), UfM policy contents must be referred to because they are necessary in order to assess possible risks and benefits the UfM carries with it for the Arab partners as well as for overall Euro-Arab relations.

*Actors: The Re-governmentalization of Euro-Arab Relations*

Among the most striking features of the UfM is undeniably that the agreed upon institutional arrangement takes the form of an inter-governmental body or even an international organization (see Johansson-Nogués, this collection). With two heads of state as co-presidents, two secretary-generals and six deputy secretary-generals selected by the seconding countries' governments, a Joint Permanent Committee (JPM) drawn from the so-called 'senior officials', all new UfM bodies that have been created or are in the process of being established are of a decidedly (inter-)governmentalist character (Aliboni and Ammor, 2009). Rather than representing the *societies* of their member countries in a broader sense, the UfM's institutions almost exclusively represent their respective governments, or rather: *heads of state* in the majority of Arab cases where these do not form part of the respective governments. While this arguably poses no serious problem in the case of the (democratic) European UfM member countries, it certainly does so in the case of the (authoritarian) Arab partner countries because here, governments do not necessarily represent the will of their populations; quite the contrary, many Arab governments and heads of state are ruling over populations who are deeply disenchanted with their autocratic leaderships.

This is remarkable because it stands in contrast to the previous arrangements of the EMP/ENP framework. Back in 1995, the EMP was founded based on the idea of three baskets, one of which was explicitly devoted to co-operation on the societal and cultural levels. The 1995 Barcelona Declaration itself stated the importance of 'support for democratic institutions and for the strengthening of the rule of law and civil society', and explicitly acknowledged 'the essential contribution civil society can make in the process of development' (Euro-Mediterranean Conference, 1995). Later developments, such as the establishment of the Anna Lindh Foundation in Alexandria and some work that evolved under the Euro-Mediterranean Study Commission, added substance to this emphasis on 'decentralized' (read: non-governmental) forms of co-operation. True, even under the older EMP framework, in actual practice these more openly political issues received far less attention from both European and Arab policy makers than Arab intellectuals and European scholars would have liked to see, but rhetorically at least they clearly figured more prominently in the official discourse than they do today.[2] Even today, the Commission still assumes that the 2004 adoption of the ENP framework for the southern Mediterranean partners was based on 'a mutual commitment to common values (democracy and human rights, rule of law, good governance...)', and 'the level of ambition of the relationship depends on the extent to which these values are shared' (European Commission, 2010).

Under the UfM, the contradictions persist. On the one hand, the 2008 'Joint Declaration of the Paris Summit for the Mediterranean', the inaugural document of the UfM, speaks of a – then already somewhat illusionary – 'shared political will' to turn the Mediterranean region 'into an area of peace, *democracy*, cooperation and prosperity' (Union for the Mediterranean, 2008; emphasis added). On the other hand, reference to those norms and values which figure so prominently in how Europe has come to portray itself remain even more on the margins than in both the EMP and ENP frameworks. While the former frameworks saw the establishment of the Euro-Mediterranean Parliamentary Assembly, the Commission-supported Euro-Mediterranean Human Rights Network and its Foundation and the European Instrument for Democracy and Human Rights, emphasis is now placed on joint projects as opposed to larger-scale and more macro-level programmes. Mutual interests in technical co-operation seem to deliberately gain in importance in relation to political areas of dissent due to regime divergence. While the UfM does not, of course, supersede the ENP, it nevertheless supersedes the EMP. The UfM does not contain those earlier ambitions of a common regional space that would incorporate respect for basic rights and freedoms.

What does that mean for the Arab partners? In a nutshell, societal forces in the Arab world will most likely look upon the UfM as yet another instance where Europe and its governments side with the much-hated Arab regimes for strategic reasons while ignoring the Arab peoples' quest for self-determination and greater freedom from oppression, both by their own governments as well as by external powers. Of course, this latter point may cause a frown among Europeans, but the feeling of being dominated by external forces is a deep-rooted perception within Arab societies that must be taken seriously if one seeks to understand prevailing reactions and attitudes towards European (or other foreign) policy initiatives (see also Moïsy, 2009: 56ff.).[3]

By contrast, Arab decision makers increasingly realize that the practical exclusion of societal forces as autonomous actors from the UfM framework is in their best interest. Struggles with the big EU donor over how far voluntary organizations will benefit from co-operation schemes will diminish to the extent that co-operation under the umbrella of the UfM, which hardly seems to know societal actors, gradually replaces schemes agreed upon within the former EMP/ENP framework under which participation of non-governmental actors figured more prominently. It is in this light that the first Arab proposals for projects within the priority areas have been submitted, indicating a change in general roles from laggards or veto-players towards hesitant supporters. In this respect – and in contrast to their stance with regard to the Arab–Israeli conflict – at least some Arab regimes have come to adopt roles of fence-sitters or cautious supporters, albeit with virtually no investment in resources, let alone for a 'common good'.

Therefore, the overall picture that derives from the UfM in relation to the key actors is one of re-governmentalization of Euro-Mediterranean relations. One could even ask whether the UfM's exclusive focus on governmental actors might contradict the very spirit of the Barcelona Declaration which saw a much more explicit role for non-state actors than does the UfM. On the other hand, this might well have been an important trigger enabling the EU to overcome initial Arab resistance.

*Institutions: The Chance for Arab Countries to become Veto-players*

The focus on regime actors outlined above comes along with an effort to generate greater co-ownership by southern partners than was evident within the EMP/ENP framework.[4] Looked at from a European perspective, the key feature of this 'inter-governmentalist' institutional design and its key marker in comparison to the EMP as well as the ENP framework may be the question of the role of the European Commission and the struggles between France and the European opponents of such an inter-governmentalization of Euro-Mediterranean relations under the UfM that engaged in efforts to Europeanize it.

By contrast, from the perspective of development policies, it seems that the core principle upheld in international development co-operation is more likely to be achieved within the institutional arrangement of the UfM than it could have been before: To ensure partners' *ownership* of programmes and projects, and thereby create a higher degree of responsibility on the partners' side in the implementation and sustenance of co-operation measures (and thus move away from earlier, more paternalistic conceptions of 'aid' towards 'true' partnerships). In fact, the MPCs not only continue to be de facto able to block European-devised Mediterranean policies, but their role has been upgraded so that they become recognized veto-players on an official, i.e. institutional level.

Even though the magnitude may not be assessed with certainty, this trait of the UfM's institutional arrangement is likely to have effects on policy choices even beyond the current priority areas. Since, according to the rules of the 'UfM game', decisions have to be taken in consensus between the two co-presidents and within the Secretariat, effective policies will only be devised (and, more importantly, implemented!) in fields where political commonalities exist and views on both sides of the Mediterranean converge. Such activities will therefore be limited to issues that are considered non-sensitive to both sides.

With respect to the Arab regimes, it is the consensus today that *the* key political priority is power maintenance (Albrecht and Schlumberger, 2004; Heydemann, 2007; Schlumberger, 2007). From an Arab perspective, this means that any measure that could threaten Arab rulers' firm grip on autocratic power will certainly not find the consent of at least one co-president. De facto, then, both co-presidents are in the position of potential veto-players with respect to any suggestion of the other side. They have the power to effectively block any policy proposal by the other which may impact negatively on topics that are deemed of vital importance to either side.

Therefore, the UfM cannot become a vehicle for any sort of meaningful political reform. By contrast, institutional reform has created an instrument for Arab regimes that makes it yet easier to avoid any reform of domestic autocratic governance.

*Policies: Two Dimensions of 'De-politicization'*

The 'de-politiciziation' of EU foreign policies towards the Mediterranean were one of the French objectives: the UfM and its activities were meant to be more immune from the Middle Eastern context in that they were assumed to be less prone to politically induced failures originating from a deterioration in Arab–Israeli relations. This contradicts Arab preferences at least in part, namely as regards the Arab–Israeli conflict: Arab countries would have preferred conflict resolution mechanisms to remain high on the European agenda whereas the EU seemed to have searched for avenues of co-operation that were less subject to possible negative spill-overs from an unresolved Arab–Israeli conflict than the initial EMP scheme (see Hollis, this collection). In this sense, a latent dissent seems to exist between the Arab MPCs that would like to see the Euro-Mediterranean framework as a stage on which they are able to pronounce their view on the Middle East conflict regularly before high-level European counterparts whereas Europe seems to have seen some virtue in a possible de-politicization of Euro-Mediterranean relations in order to not have possible progress on other issues impeded by a stalled regional conflict.

As Asseburg and Salem (2009: 13) state, the Arab–Israeli conflict has been 'one of the main stumbling blocks to progress towards regional cooperation, stability, and economic and political reform in the Mediterranean. It has also severely impaired confidence and trust building between the two shores of the Mediterranean'. But the solution to enhanced Euro-Mediterranean relations will hardly lie in ignoring the existence of this conflict and trying to exclude it from the agenda of the UfM.

The relevance of this tricky topic, however, can ultimately be traced back to *domestic* Arab politics. It is essential for regimes that enjoy little legitimacy at home not to open up new areas of popular dissent, and given the broad public support the Palestinian cause receives in Arab public opinion, there is a strong disincentive for Arab regimes to adopt an accommodating attitude with respect to the Arab–Israeli conflict on the international scene – a disincentive deriving from the domestic Arab polity level (on legitimacy in Arab states, see Schlumberger, 2010). But there is a different, albeit related, dimension of 'de-politicization' – a dimension which also has to do with Arab domestic politics. Arguably, this is the more important dimension or understanding of 'de-politicization' which enters Euro-Arab relations with the advent of the UfM.

On the one hand, the Arab–Israeli conflict and its global dimensions impact on the general prospects of the UfM, and namely on the question of the willingness of Arab partners to engage in regional policy dialogue and projects. On the other hand, it is precisely because of the Arab–Israeli conflict's global dimensions that European options to assume a more active role, possibly one of 'honest brokerage', remain limited. By contrast, while European options as regards political reform *within* individual Arab partner countries are limited as well, they are far less prone to influences beyond Euro-Arab relations than is the case with the EU's role in the Middle East conflict. This topic neither involves that conflict directly, nor necessarily transatlantic considerations. European support for political reform in the Arab world can be self-determined in an autonomous manner by the EU to a much greater degree than can European policies towards the Arab–Israeli conflict. The question of European support to Arab political reform, in turn, is an extremely relevant policy issue because the political stagnation and lack of political development which has characterized the Arab world over recent decades is *the* key obstacle to enhanced economic and human development in the region (see, e.g., UNDP/AFESD, 2005), along with a range of ensuing political questions and challenges that impact directly on European political and security interests (see my introduction above). This has not only been unambiguously and boldly pointed out by the United Nations Development Programme's (UNDP's) regional reports on human development (the Arab Human Development Report [AHDR] series), but also by a broad consensus of scholars specialized in Middle East politics and political economy.

Let me therefore take a closer look at what impact the renovation of the overall Euro-Mediterranean framework under the UfM may have on the question of the EU as an effective supporter of processes of political reform in the Arab world. The UfM from its earliest stages has been dubbed a 'project of projects' or a 'union of projects' in which, as hinted at above, concrete projects are given priority over large and 'high' policies such as conflict resolution or democracy assistance. The six priority projects that have been identified for future UfM activities[5] display a high degree of similarity in terms of their technical (as opposed to political) nature to development schemes of the 1960s when international development co-operation was inspired by classical modernization theory's belief in the primacy of economic growth which later on would somehow trigger democracy.

What is clear from comparing the UfM's design to the EMP/ENP policy framework – apart from its institutional and actor-related specifics – is that political conditionality seems to have been given up.[6] Not only is the principle of conditionality unknown in the UfM proper, but also policy contents have been technicized to the extent that political reform no longer seems to be part of Europe's ambitions in its relations with the MPCs. In this sense, the transfer from EMP to UfM may not be all that smooth and harmonious. While struggles behind the scenes (and, in fact, also on stage) have been discussed in other contributions, certain questions as regards the compatibility of the old framework with the new can also be asked with respect to policy contents. The technicization of relations inherent in the UfM, Schwarzer (2009) holds, 'contradicts the aims of the previous Barcelona Process, which understood Europe's involvement in the region primarily as a contribution to transforming the economic and political systems in the southern Mediterranean area'. Indeed, combating pollution in the Mediterranean sea, schemes for co-operation in infrastructure, energy, and even in trade and private sector development surely have the potential to result in quicker consensus among the dual structure than in the past, but do not address the overarching issues necessary for political, human and economic development. Schwarzer (2009) even goes as far as to allude to potential risks for Europe inherent in such a de-politicized approach:

> the de-politicisation of the EU's Mediterranean policy taking place within the project-focused cooperation, coupled with an insufficiently developed involvement of civil society in the southern Mediterranean states, is unlikely to promote transformation, and may even reduce transformative potential. A policy that does not seriously support transformation processes, however, will do no favours to the EU member states' medium and long-term security interests.

**Two Scenarios for Future Euro-Arab Relations within the UfM Framework**

The UfM will not transform Euro-Mediterranean relations altogether. With regard to Euro-Arab relations in particular, it will certainly change some of the previously existing patterns of interaction, but it will do so to a limited extent only. This is so because, on the one hand, the institutional arrangements that are being installed under the umbrella of the UfM do not supersede and replace the previous frameworks entirely, but can rather be seen as a complement that will contribute to a re-arrangement of the previous policy frame in some respects, but not to its demise. It has been made clear in all relevant official deliberations that bilateral association agreements will remain in place and valid. This has been pointed out and explained in various contributions on the UfM and therefore need not be repeated at length (Aliboni and Ammor, 2009).

The second reason that limits the degree to which the UfM will alter Euro-Arab relations lies in the latter's geostrategic importance not only as Europe's neighbours. The Arab Mediterranean countries, of course, are part of the family of Arab nations which represent not only a relatively homogeneous group of countries as regards a joint linguistic, cultural and religious heritage, but, more importantly, which represents the world's largest reservoir of exportable fossil fuels – and these, in turn, will remain indispensable for the OECD member countries' economies despite the gradual rise of renewable energies for decades to come (Schlumberger, 2006). Moreover, with the rapid increase in new great powers' thirst for oil, the EU has no alternative but to entertain as good as possible political relations with the providers of what makes their economies run.

For all that, the UfM will alter existing patterns of Europe's relations to the Arab countries to a limited extent only. This does not mean that the UfM has no impact at all on future Euro-Arab relations. Let me therefore propose two scenarios pointing to the different directions these relations may take in the future due to the influence of the UfM. First, I sketch out the possibility of future Euro-Arab relations evolving along the officially proclaimed logic that, as the public has been told by the French administration, underlies the UfM, and which could be dubbed the 'Schuman–Sarkozy scenario'. Second, I propose an alternative scenario which might be termed the 'pessimist picture' or the 'realist scenario' in the sense that it is based less on a functionalist logic compatible with a social-constructivist understanding of EU foreign policies than on a more traditional or 'realist' understanding of the nature of international relations (see Holden, this collection).

*The 'Schuman–Sarkozy Scenario'*

As has been hinted at above, one of the core features of the UfM is its focus on technical projects and on the introduction of possibly more flexible possibilities of ad hoc co-operation in non-sensitive areas where consent is relatively easier to achieve. This comes along with an obvious 'lack of mention of strategic issues of politics', as Emerson (2008: 11) puts it, based on the EU's 'extreme caution over issues of desirable democratic political reforms and respect for human rights' (Emerson, 2008: 11). Presumably, however, this 'extreme caution' is not (only?) rooted in what some might call European cowardice or accommodativeness (i.e. shying away from the more complicated and conflictual issues in the partnership), but in a so-called 'functional logic'. As Balfour and Schmid (2008: 4) point out, 'according to the French, the "union of projects" approach is inspired by EU founding father Jean Monnet's functionalist method and his "solidarités de fait" (solidarity created by actually achieved facts)'.[7] In brief, this logic holds that the modest creation of functional links will over time result in deeper integration, similar, in one sense, to interdependence theory which argues that increasing levels of (economic) interaction will lead to increasingly non-violent modes of conflict regulation.

It cannot be denied that this 'functional logic' has some attractiveness to it: Translating this logic into a scenario for the future, functional co-operation in (some few) areas of truly shared interests will over time lead to increased levels of interaction, to a widening of the areas of co-operation, to deeper Euro-Mediterranean integration, and will ultimately – after the political systems north and south of the Mediterranean have converged significantly enough due to this dynamic of functional co-operation – one day even convince Arab rulers to cede power in favour of their hitherto disenfranchised populations. Normalization of relations with Israel and democratization, then, will not come about tomorrow, but they can and will come about. There is thus no need to aggressively exert pressure for quicker or more meaningful political reform on the MPCs; rather, this would be detrimental to the development of an overall positive political climate and an evolving spirit of co-operation. 'Core policies, such as conflict resolution or political reform and democratisation, are being excluded in favour of small-scale cooperation in key areas of common interest, with the aim of improving the overall political climate in the region' (Balfour and Schmid, 2008:3). What adds to the plausibility of this scenario is that while the divergence of political regimes around the Mediterranean may seem great today, levels of trust and confidence were certainly no higher between, say, France and Germany in the aftermath of World War II when Schuman delivered his declaration than those that exist today between the Arab world and the EU and its members after 15 years of the Euro-Mediterranean partnership. In this logic, then, there is not only no need to pursue active peace policies or democratization efforts, but these could even have a negative impact on the evolution of the overall UfM.

*The 'Realist Scenario'*

Let there be no doubt that this author would gladly welcome a future development of Euro-Arab relations along the lines of the 'Schuman–Sarkozy scenario' as outlined above. But there is an alternative scenario of future Euro-Arab relations which differs decisively from the one just presented.

This second or 'realist' scenario holds that even a power that might consider itself 'normative' by nature will eventually not run the risk of devising (let alone implementing) policies that contradict its own material interests and preferred policy outcomes in international relations. It will therefore prefer pragmatism and self-interest over a normative vision, even if that latter was grounded in a functional logic. The second scenario therefore starts from revisiting the agreed upon frame of the UfM with its three core dimensions of actors, institutions and policies as discussed above and, from there, concludes on the likelihood of major alterations the new political frame might bring about or not.

The three core dimensions of the UfM reveal clearly that the European side has largely given up on previously established larger and normatively based foreign policy goals in its relations with the Arab world. While core values such as the rule of law, respect for human rights, or participatory and inclusive modes of governance have – at least on a rhetorical level – figured prominently in the EMP framework established at Barcelona in 1995 and also in the Neighbourhood Policy pursued since 2004, these hardly appear at all in the renovation of Euro-Arab policies under the UfM. Quite the opposite is the case

1. With regard to *actors*, societal forces from the Arab world have virtually been excluded and their inconvenient voices seem less audible in the UfM. This represents a political priority which Arab regimes, through elaborate mechanisms of co-optation and repression in neopatrimonial contexts (Albrecht and Schlumberger, 2004), had already effectively implemented on national levels, but which have now been transferred to the international level of Euro-Arab relations through the UfM.
2. It is unlikely that despite this exclusion of potentially deviant voices policies will be suggested within the frame of the UfM that could be regarded as inopportune to the prime Arab political priority of authoritarian regime maintenance. The new *institutional arrangements* elevate Arab regimes to institutionalized veto-players and allow the Arab side to block any unwelcome political initiatives more effectively than before, even without Arab regimes having to agree on paper first and circumventing agreements in practice later. Under the UfM they can block any initiatives that run counter to their desire to cling to power and even prevent them from being formulated as Mediterranean policies. This can arguably be considered a second safeguard mechanism towards the maintenance and consolidation of authoritarian rule in addition to the first aspect. While co-ownership in theory is nice to have, this is only so in practice if the respective partners agree on joint political preferences.
3. *Policy contents* as reflected in the UfM's prioritized issue areas represent a near total de-politicization of co-operation. This makes it less likely that Arab regimes' political priorities of power maintenance are addressed. De facto, then, political conditionality – an instrument never used, but theoretically existing under the EMP scheme and within the ENP – has been abandoned in the new framework of the UfM.

The UfM in this sense, then, represents a triple victory of Arab political preferences – and a clear step back from earlier EU policies that not always exerted high pressure for norm-compliance or political reform, but which nevertheless made an explicit claim that there could be a long-term European policy goal related in one way or another to what are regarded in Europe as core norms and values of politics. The disappearance of these norm-based policy goals in the UfM certainly does not only represent a European naivety in foreign policy making. Rather, it seems that European decision makers have come to accept the fact that insisting on political goals (as opposed to micro- or meso-level technocratic goals) might not serve European interests best: If pursued seriously, this might in some cases even lead to ruptures in Europe's transatlantic relations (because of strategic US support to key Arab regimes), not to speak of the potentially devastating impacts systemic political transformations might have for regional stability – the latter being a necessary ingredient for European energy security, or for safeguarding markets that have been opened up not least through the Euro-Med economic co-operation schemes for the export of European industrial (and financial) products.

In other words, the UfM may not only be considered a triple victory for Arab authoritarian regimes, but may also be read as a convergence of interests by European and Arab political elites around the political status quo. As Kausch and Youngs (2009: 967) argue: 'the EU has moved further and further away from seeking a "ring of well-governed states" on its southern edge towards seeking a "ring of *firmly* governed states"'. Note, however, that this is not per se illegitimate – after all, European policy makers are supposed to pursue European political interests. However, the more norms and values are openly disregarded in both European foreign policy design and policy implementation, the less convincing it becomes to cover up real and material interests in a public political rhetoric based on such norms and democratic values vis-à-vis the European public.

The official Schuman–Sarkozy scenario, then, seems like a rather helpless effort at such public rhetoric since it neglects the key ingredient of any type of deeper integration, which is *political regime convergence*. Even during the rivalry between France and Germany after World War II, both countries were rooted in democratic constitutions and the issue was about reconciliation rather than about a clash over fundamental norms of how the political process can legitimately be conducted. The issue in the relations between a democratic Europe and authoritarian Arab regimes, by contrast, is that functional co-operation will not serve as a vehicle to regime convergence so that these differences can be overcome. This is because they are differences in kind, not in degree, and because the maintenance of these differences is made a precondition for co-operation by one side.

Realistically, then, if Europe does want co-operation – and in the absence of military options for forced regime overthrows – it seems to serve the EU's own economic and security interests best to engage in such co-operation under given political constellations. At any rate, the UfM demonstrates forcefully that efforts at altering or transforming contradictory political regime constellations between north and south of the Mediterranean are no longer on the European agenda, if ever they have been. With Seeberg (2010: 291), then, 'we can speak of cooperation between two uneven partners where the dominant part realizes that *realpolitik* is what can be pursued in the Middle East'. The only question, however, seems to be about who is the dominant part in this game of Euro-Arab relations: Bearing in mind the triple victory for Arab political priorities, Europe does not give the impression of being the dominant player in this game, even though certainly the one who invests the lion's share in terms of resources into the partnership. However, this could – from a realist perspective – relatively easily be explained by the fact that in Euro-Arab relations, the potential gains for Europe in the realms of security and the economy are much larger than those for the Arab side. I concur, then, with Peter Seeberg that the EU is to be viewed as 'a realist actor in normative clothes' (Seeberg, 2009: 81), and, to stay within the analogy, that the UfM strips off more and more of these clothes from the materialist, self-interested, and pragmatist body of European foreign relations towards its southern neighbours. The fundamentally erroneous notion of a qualitative difference of the EU's foreign policy behaviour as more norm-based in comparison to other actors on the international scene such as the USA, India or China thus becomes increasingly difficult to uphold.

**Conclusions**

Obviously, the scenario based on the functional logic of Europe's own integration does not take into account the qualitative differences between the European integration process among democracies on the one hand and the very heterogeneous group of democracies and autocracies around the Mediterranean on the other. Power distribution within national political systems and the resulting modes of governance, as is well established, have an enormous influence on prospects for both integration and development. The idea of replicating around the Mediterranean today what worked over past decades among European democracies is wishful thinking and contradicts not only everything we know today about the impact of European efforts at promoting democracy in its near abroad without offering full membership (cf. inter alia Schimmelfennig and Scholz, 2008). What is more, even when offering full membership, the EU's effort to promote democratic rule elsewhere has only been successful if and when political conditionality was employed. As Comelli (2009: 161) notes correctly: 'If conditionality – together with socialisation – was the most effective method through which the EU successfully transformed Central and Eastern countries, it follows that without the application of conditionality, the reforms envisaged by the ENP are doomed to fail.' The UfM threatens to eliminate political conditionality from Euro-Mediterranean relations altogether or at least to greatly reduce it. Therefore, the second scenario, which is based on the assumption that the EU has its own policy priorities (preventing terrorism, combating illegal migration and transnational organized crime, etc.), which it pursues independently of whether it spreads its own norms and values, is much more likely to unfold over the next decade.

If the purpose of the present contribution was a brief and concise prediction of the UfM's impact on the future of Euro-Arab relations, I would stress two points. First, its impact will be of an incremental rather than transformative nature, all the more so since the former rhetoric of political reform and democratization was never seriously given priority by the EU in the past and therefore the de-politicization inherent in the UfM's design will not result in *dramatic* shifts in actually implemented policies. Second, and notwithstanding the first point, the shift in the framework as regards the three dimensions under consideration in this contribution (actors, institutions and policies) carries in it an inherent tendency towards a stabilization of authoritarian modes of governance as well as towards de-development. This is so even though in the short term the UfM may benefit from renewed activism and maybe consent between the EU and its Arab neighbours may be reached somewhat more easily in the apolitical priority areas for future co-operation. But if it is correct that 'the EMP can only be reinvigorated and maintain its relevance into the future if it turns its *potential* acquis into *specific actions* designed to create a Euro-Mediterranean community of democratic states' (EuroMeSCo, 2005: 9), then the UfM does not have the potential to spur deeper integration in the long run, or to provide a substantial impetus for sustainable economic and human development within the Arab world: both depend on meaningful Arab political reform on the domestic scene.

Ironically, therefore, the UfM inherently contradicts the declared policy goals (long-term economic development; economic and political integration through functional co-operation) its designers say they wish to achieve. The UfM not only represents a European capitulation to an intractable Arab–Israeli conflict but also to the challenge of fostering greater respect for human rights and more participatory governance inside the Arab partner countries. While this will be welcomed by Arab governments, they will – as initial reactions demonstrated – nevertheless remain hesitant to enter into much closer relations with the EU than they already have because of their unwillingness to normalize relations with Israel and, second, because of the fears of getting caught in too close a political network in which Europe could be able, one day, to dictate policies that might be perceived as an infringement of national sovereignty or to impact negatively on vital questions of political power distribution within their own borders.

Having said this, the above provides more than just an assessment of the future shape of Euro-Arab relations. Rather, it can also be read as a contribution to the growing counter-evidence against the essentialist notion of a 'normative power Europe' which some scholars still tend to see in the EU and its foreign policies. True, the new institutional set-up of the UfM, especially its co-Presidency, the composition of the Secretariat and the JPC drawn from both sides of the Mediterranean as well as other monthly meetings suggest that it may not necessarily be correct to conceive of Euro-Mediterranean policies as 'the most imperialist project since World War II', as argued at a 2000 conference panel on the Euro-Mediterranean Partnership initiative.[8] Yet the EU will most likely not feel comfortable with giving up substantial policy priorities merely for the sake of achieving greater MPC co-ownership.[9]

Thus, apart from the fact that the concept of 'normative power' itself still awaits a viable operationalization, available empirical evidence suggests that – no matter how we define it – Europe in its foreign policies is no more of a 'normative power' than are the US – or even China, for that matter. Rather, it seems that Europe's Mediterranean policies continue to be one of the international factors which play a role in the durability of authoritarian rule in the Arab world. In the words of Comelli (2009: 161): 'To be questioned is not only the effectiveness of the EU in transforming the economic and political systems in the Mediterranean, but also its willingness to do so.' In this light, the present contribution may also be read as part of a newly emerging literature that focuses on international factors not in democratization but in the survival of autocracies and the 'authoritarianization' of national political regimes (Burnell and Schlumberger, 2010).

## Acknowledgements

Earlier drafts of this study were discussed at LSE London, 10 May 2010, and at the World Conference of Middle East Studies, 16–20 July 2010, Barcelona. The author wishes to thank the editors of this special issue as well as Annette Jünemann and an anonymous referee. Special thanks go, again, to Federica Bicchi and Richard Gillespie for detailed and critical but very helpful comments, as well as careful edits.

## Notes

[1] See Bicchi's introduction as well as Hollis' contribution to this collection for details. For details on Arab fears for normalization, see especially Khatib (2010).

[2] For discussions of the earlier Barcelona Process under the architecture of the EMP/ENP with a focus on its impact on the Arab countries, see Schlumberger (2000) or Kienle (1998).

[3] To be sure, this rhetorical emphasis of the need for political convergence around democratic governance never translated in actual practice the way it was initially intended – and, of course, the 'Euro-Mediterranean Community of Democratic States' which 'Barcelona Plus' Euromesco Report of 2005 had in its title will not materialize any time soon. This report summarized the situation quite clearly: 'Progress towards democracy has fallen short of original expectations, and thus the degree of political convergence on which integration is predicated has failed to materialise' (EuroMeSCo, 2005: 22).

[4] For details of the UfM's institutions, see Johansson-Nogués (this collection).

[5] For more detailed discussions of the priority areas see Darbouche (this collection) and Hunt (this collection). The 'Business Development Initiative' (priority no. six) reads like a remnant of the failed neoliberal approach the IMF and the World Bank have pursued for over 20 years in their respective macro-economic reform and structural adjustment programmes (ERSAP), which in the Middle East and North Africa have failed to produce desired outcomes mainly because the legal and political environment for private business, above all the structural absence of the rule of law and contract security as well as impartial arbitration mechanisms and effective enforcement of anti-trust legislation, had been thoroughly ignored (Heydemann, 2004; Schlumberger, 2008).

[6] While it is true that the bilateral Association Agreements which contain a political conditionality clause referring to human rights and democracy remain in place, the routine violations of this by Arab regimes have never been raised by the EU. With policy contents becoming less contentious and more technicized, the gap between actually implemented policies on the one hand and political principles the EU would like to uphold on the other is bound to widen and the likelihood of the said clause gaining practical relevance is decreasing.

[7] Back in 1950, French foreign minister Robert Schuman, in a speech written together with his advisor and friend Jean Monnet, had put forth the first proposal for a 'European federation' based on a union between France and Germany (plus others) for the joint exploitation of coal and steel. However, with the symbolic issues of coal and steel (both essential for military industry) he clearly envisaged the establishment of a 'European federation' that would one day be a guarantor for peace in Europe (Schuman, 1950).

[8] The quote is borrowed from a political science colleague and former president of the British Society for Middle Eastern Studies.

[9] Moreover, a proposal by Arab League Secretary General Amr Mousa that was tabled at the Arab summit held in Sirte, Libya, in March 2010 (Libyaonline.com, 2010) has gained international attention lately. The Arab League recently released plans to establish a 'Union of Arab Neighbours' as a body that is to co-operate with other regional organizations such as the EU or the African Union as well as with larger neighbouring countries of the Arab world (Abu Husain, 2010). Whether this uncoordinated, but almost simultaneous announcement of two inter-governmental unions really represents an Arab effort to counterbalance potential negative impacts of the UfM for the Arab countries does not matter here; at any rate, it demonstrates that Arab countries are cautious not to rely too heavily on the promises of European policy initiatives. If Europe's normative radiance was as great as some of us assume, this would certainly not have been a necessary development – in other words, the level of trust between actors that have been Mediterranean partners for more than a decade in a 'thick institutional context', as Bicchi correctly points out in her introduction to this collection, should be assumed to be high enough for these partners to closely co-ordinate two similar policy initiatives which clearly overlap in both geographical scope and in the goals they seek to achieve.

# References

Abu Husain, S. (2010) A conversation with Arab League SecGen Amr Musa, *ash-Sharq al-Awsat*. Available at http://www.aawsat.com/english/news.asp?section=3&id=20664 (accessed 26 July 2010).

Albrecht, H. & Schlumberger, O. (2004) 'Waiting for Godot': regime change without democratization in the Middle East, *International Political Science Review*, 25(4), pp. 372–391.

Aliboni, R. & Ammor, F. (2009) Under the shadow of 'Barcelona': from the EMP to the Union for the Mediterranean, *EuroMeSCo Paper*, no. 77 (January).

Asseburg, M. & Salem, P. (2009) No Euro-Mediterranean community without peace, *10 Papers for Barcelona 2010*, No. 1 (September) (Paris: European Union Institute for Security Studies).

Balfour, R. & Schmid, D. (2008) Union for the Mediterranean, disunity for the EU? *Policy Brief February 2008* (Brussels: European Policy Centre).

Burnell, P. & Schlumberger, O. (2010) Promoting democracy – promoting autocracy? International politics and national political regimes, *Contemporary Politics*, 16(1), pp. 1–15.

Comelli, M. (2009) Conclusions, in: M. Comelli, A. Eralp & Ç. Üstün (Eds) *The European Neighbourhood Policy and the Southern Mediterranean. Drawing Lessons from the Enlargement*, pp. 157–165 (Ankara: METU Press).

Emerson, M. (2008) Making sense of Sarkozy's Union for the Mediterranean, *CEPS Policy Brief No. 177* (March) (Brussels: Centre for European Policy Studies).

Euro-Mediterranean Conference (1995) *Barcelona Declaration adopted at the Euro-Mediterranean Conference 27–28 November 1995*. Available at http://trade.ec.europa.eu/doclib/docs/2005/july/tradoc_124236.pdf (accessed 20 October 2010).

EuroMeSCo (Ed.) (2005) *Barcelona Plus. Towards a Euro-Mediterranean Community of Democratic States* (Lisbon: Institute of Strategic and International Studies). Available at http://www.euromesco.net/media/barcelonaplus_en_fin.pdf (accessed 20 October 2010).

European Commission (2010) The policy: what is the European Neighbourhood policy? Available at http://ec.europa.eu/world/enp/policy_en.htm (accessed 12 October 2010).

Henry, C. M. & Springborg, R. (2001) *Globalization and the Politics of Development in the Middle East* (New York: Cambridge University Press).

Heydemann, S. (Ed.) (2004) *Networks of Privilege in the Middle East. The Politics of Economic Reform Revisited* (New York: Palgrave).

Heydemann, S. (2007) Upgrading authoritarianism in the Arab world, *Saban Center Analysis Paper*, No. 13 (October) (Washington, DC: Brookings Institution). Available at http://www.brookings.edu/papers/2007/10arabworld.aspx (accessed 20 October 2010).

Kausch, K. & Youngs, R. (2009) The end of the Euro-Mediterranean vision, *International Affairs*, 85(5), pp. 963–975.

Katz, J. (2008) Proposed Mediterranean alliance faces resistance. Available at http://www.findingdulcinea.com/news/international/May-June-08/Proposed-Mediterranean-Alliance-Faces-Resistance.html#3 (accessed 31 October 2010).

Khatib, K. (2010) The Union for the Mediterranean: views from the southern shores, *The International Spectator*, 45(3), pp. 41–50.

Kienle, E. (1998) More than a response to Islamism: the political deliberalization of Egypt in the 1990s, *Middle East Journal*, 52(2), pp. 219–235.

Libyaonline.com (2010) Arab summit twenty-second leader of the revolution headed by close of its meeting held in Sert in a public hearing [sic.]. Available at http://www.libyaonline.com/news/details.php?id=12855 (accessed 20 October 2010).

Manners, I. (2002) Normative power Europe: a contradiction in terms? *Journal of Common Market Studies*, 40(2), pp. 325–358.

Moïsy, D. (2009) *The Geopolitics of Emotion: How Cultures of Fear, Humiliation and Hope Are Re-Shaping the World* (New York: Doubleday).

Pace, M. (2007) Norm-shifting from EMP to ENP: the EU as a norm entrepreneur in the South? *a Cambridge Review of International Affairs*, 20(4), pp. 659–675.

Reuters Africa (2008) Libya and Turkey object the Union for the Mediterranean, http://af.reuters.com/news/country/?type=libyaNews (accessed 30 December 2009).

Schimmelfennig, F. & Scholz, H. (2008) EU democracy promotion in the European neighbourhood: political conditionality, economic development and transnational exchange, *European Union Politics*, 9(2), pp. 187–215.

Schlumberger, O. (2000) Arab political economy and the European Union's Mediterranean policy: what prospects for development? *New Political Economy*, 5(2), pp. 247–68.Schlumberger, O. (2006) Rents, reform and authoritarianism in the Arab Middle East, *International Politics and Society*, 2, pp. 57–72.

Schlumberger, O. (2007) *Debating Arab Authoritarianism. Dynamics and Durability in Nondemocratic Regimes* (Stanford, CA: Stanford University Press).

Schlumberger, O. (2008) Structural reform, economic order, and development: patrimonial capitalism, *Review of International Political Economy*, 15(4), pp. 622–649.

Schlumberger, O. (2010) Opening old bottles in search of new wine: on non-democratic legitimacy in the Middle East, *Middle East Critique*, 19(3), pp. 233–250.

Schuman, R. (1950) Déclaration du 9 mai 1950. Available at: http://www.robert-schuman.eu/declaration_9mai.php (accessed 25 October 2010).

Schwarzer, D. (2009) Future of the Mediterranean Union. growing challenges, lack of political will, *Qantara*. Available at http://en.qantara.de/webcom/show_article.php/_c-476/_nr-1152/i.html (accessed 20 October 2010).

Seeberg, P. (2010) Union for the Mediterranean – pragmatic multilateralism and the depoliticization of EU–Middle Eastern Relations, *Middle East Critique*, 19(3), pp. 287–302.

Seeberg, P. (2009) The EU as a realist actor in normative clothes: EU democracy promotion in Lebanon and the European Neighbourhood Policy, *Democratization*, 16(1), pp. 81–99.

Union for the Mediterranean (2008) *Joint Declaration of the Paris Summit for the Mediterranean*. Available at http://www.emuni.si/Files//Dokumenti%20PDF/Joint_declaration_of_the_Paris_summit_for_the_Mediterranean-EN.pdf (accessed 30 October 2010).

United Nations Development Program (UNDP)/Arab Fund for Economic and Social Development (AFESD) (2005) *Arab Human Development Report 2004: Towards Freedom in the Arab World* (New York: UNDP-Regional Program for the Arab States, Iran and Turkey).

Vucheva, E. (2008) Arab countries complicate Med Union plan, http://euobserver.com/9/26293 (accessed 30 October 2010).

Winkler, P. (2008) Was bringt die Mittelmeerunion, *Neue Zürcher Zeitung*, 14 March. Available at http://www.eurotopics.net/de/presseschau/archiv/article/ARTICLE25463-Was-bringt-die-Mittelmeerunion (accessed 26 July 2009).

# A New Beginning? Does the Union for the Mediterranean Herald a New Functionalist Approach to Co-operation in the Region?

PATRICK HOLDEN
Politics and International Relations Group, Plymouth Business School, University of Plymouth, UK

ABSTRACT *The UfM is building on an already extensive policy acquis to support regionalism and regionalization. Regionalism is understood here as the elaboration of formal trade and economic integration agreements between governments; whereas regionalization refers to the development of relations beyond the governmental level (trade, FDI, joint ventures between businesses; and links between other elements of society). These interconnected processes play a major role in the EU's vision for the region. The EMP envisages bilateral and multilateral free trade agreements while the ENP encourages further regulatory harmonization. There has also been strong proactive support for regionalization with aid and co-operation policies to foster co-operation between businesses, technocrats, social organizations and cultural and educational institutions. Apart from developing common institutions, the UfM is less directly focused on economic regionalism, and more on a form of regionalization via support for joint projects in core sectors. There appears to be a kind of functionalist ethos of flexible, bottom-up project formation. This poses important questions. Does the UfM signal a break with the essentially neo-liberal thrust of EU policy and a move to a more functionalist and interventionist approach to supporting development and co-operation? What is the potential for this to foster regionalization and development in the area? From the information available, there is little reason to believe that the UfM will be more successful than previous efforts at encouraging regionalization. In terms of our understanding of it, we must bear in mind that the UfM does not replace the EU's core 'trade and integration' policy. Therefore it is more of an addition to EU policy than a change of direction.*

At first glance, the Union for the Mediterranean (UfM) appears to offer a new approach to collaborative regionalism in the Mediterranean. The Sarkozy vision, agreed at Paris in July 2008, involves some distinct innovations from the standard EU template. Rather than trade-related measures and liberal reforms, the core activities envisaged are large transnational projects in key socio-economic areas.

The supranational Secretariat and co-Presidency appear to offer a renewed form of partnership and a different *modus operandi*. However the UfM is only the latest in 40 years of European initiatives to develop co-operation in the region. In particular the major policies and institutions of the Euro-Med Partnership and the European Neighbourhood Policy are still very much in effect. Therefore, although the UfM is still in its infancy, it is worth analysing how it compares with these EU approaches, and how it will interact with their policies and processes. The focus here will be on the economic dimension and the overarching question is whether the UfM represents a significant change in EU policy and Euro-Mediterranean relations?

The analytical framework of this study is based on the concepts of regionalism and regionalization. Regionalism is understood as the elaboration of agreed rules and collective institutions by the governments of a region; generally these are formal trade and economic integration agreements but could extend to other areas. On the other hand, regionalization refers to the development of relations beyond the governmental level including trade, FDI/joint ventures between businesses; and links between other elements of society (Breslin and Higgott, 2000). Another way of understanding this, with regard to trade policy, is that regionalism 'is the idea that countries should preferentially promote trade with countries of the same region' while regionalization refers to a 'natural' pattern of trade and interaction stemming from geographical, cultural and economic factors (Tovias, 2000: 326). Of course the two concepts are interconnected and the term 'regional integration' is often used to refer to both processes (formal agreements and actual economic/societal interrelations). Clearly, EU Mediterranean policy has been attempting to engender both processes. However, the concepts are worth distinguishing between and it is possible to have either one without the other, for example regionalism (formal agreements and institutions) may not lead to actual regionalization. While regionalism of different types can be seen as an end in itself, regionalization is the deeper objective of EU Mediterranean policy: the hope being that increased interaction between economic and social forces will promote peace and development.

How does this relate to the UfM? It is clearly a form of regionalism in that it is a high-profile intergovernmental agreement. However the economic aspect of it is supposed to be more concerned with direct joint action than with formal inter-state integration. Therefore it could be conceived of as a more direct focus on promoting regionalization. What is the potential for this, and how may we divine the impact of the UfM approach on regionalization in the area? A second, related, question focuses more directly on the nature of the EU's policy. The ethos of the UfM implies a strategy to promote peace and development based on flexible co-operative projects (involving the public and private sectors) in key socio-economic areas. In this respect, the UfM appears to be in line with classical functionalist thinking: the belief that flexible, transnational institutions with clear practical functions could ameliorate international economic stagnation and security tensions (Mitrany, 1944). On the other hand, the more politicized aspects of the UfM are clearly at odds with functionalism norms and methods. In brief, if we theorize the official discourse of the UfM there is some evidence to suggest that it proposes a new, more functionalist approach to promoting development and regionalization, which can be contrasted with free trade-based regionalism. (As discussed below, although the original functionalist thesis was not anti-trade, it posited a strong interventionist role for international public agencies.) The pertinent question here is whether the UfM signals a break with the hitherto essentially neo-liberal thrust of EU policy which has been based overwhelmingly on free trade and market power?

The contribution will begin by considering different theoretical frameworks that help us to characterize the EU's economic policies and institutional proclivities. This is necessary to help clarify whether the Union for the Mediterranean package is a significant departure. On a more empirical level, it will then review the core features and priorities of previous support for regionalism and regionalization. Without wishing to regurgitate all of the EU's previous policies, which have already been studied intensely, it is important to emphasize the extent of previous EU policies aimed at achieving similar objectives to the UfM. (An underlying theme here is the deeper question as to what an institution such as the EU can hope to achieve regarding bottom-up regionalization.) It then analyses the UfM in itself, within the context of these previous policies. Much of this study is necessarily based on discourse and official documents as the UfM has not 'done' very much yet. All of this should help us understand precisely what (if anything) is new and significant about the UfM in this sphere.

## The EU's Pattern of External Relations: Neo-liberal and Neo-functionalist

As implied above, the EU's model is considered here to have been essentially neo-liberal, but there are numerous caveats which must be considered. The term 'neo-liberal' is often ascribed very broadly (it tends to be used to describe almost any form of capitalism in the globalization era). Furthermore there is an inherent difficulty in characterizing the complex (often contradictory) policies of any organization in terms of 'ideal type' economic and political concepts. This is especially true of the EU, which takes complexity (and incoherence) to extremes. The nature of the EU is still very much contested with regard to whether it is a neo-liberal entity devoted to global free markets or a more statist/social democratic model (Van Der Pijl, 2006). Political perspectives within Europe may view the EU as effectively synonymous with neo-liberal globalization or as a vehicle for some form of Euro-statism. This essay assumes that in its external relations its discourse and action has been essentially neo-liberal.

Pure neo-liberal socio-economic theory advocates that economic activity should be determined by voluntary market-based interactions within the private sector; rather than centralized political efforts to guide the economy (Friedman, 2002). Free markets, according to this thesis, are infinitely more efficient in their allocation of resources and lead to more wealth creation. Neo-liberals such as Friedman and Hayek advocated an absolutely minimalist approach to political intervention and political institutions: the role of public power is merely to set the rules of the game (Friedman, 2002; Hayek, 1979). At the international level, neo-liberal theory strongly supports global free trade and economic integration, in line with the principles expressed above. Most contemporary states and international institutions are often regarded as essentially neo-liberal as they privilege the role of market forces in contrast to any kind of interventionism. However, in reality, no major state has implemented anything approaching the full neo-liberal vision in the domestic realm (as all kinds of subsidies, fiscal transfers and social safety nets persist). Polanyi's famous argument that a truly free market is incompatible with social and political stability has not been tested by national governments (Polanyi, 2001).

International organizations such as the WTO and the IMF may be better described as neo-liberal in ethos and function. Some regional entities such as the North American Free Trade Area are clearly neo-liberal, while others, such as Mercosur, are more circumspect in their posture towards free trade and global markets. Similarly the pro-free trade posture of developed states is by no means unequivocal and their approach, if it is assumed to flow from economic theory, is more akin to that of strategic trade theory (Krugman, 1986) than purely neo-liberal principles. The overall approach of powerful developing states is best described as instrumental neo-liberalism: neo-liberal principles permeate policy but they are not applied uncritically (power considerations and vested interests intervene).[1]

This notion of instrumental neo-liberalism plainly also applies to the EU. Policies such as the CAP and the structural funds bear witness to a clear desire to intervene in the free market when appropriate. In its external relations the EU has always presented itself as less rigorously free-market than the United States but the substance of its policies is not so obviously different. The major thrust of its policies is based on regional and global trade liberalization. The comprehensiveness of its trade agenda precludes any kind of strategic trade or 'infant industry' policy on the part of its partners. The core element of its aid to regional integration is technical assistance to remove obstacles to trade. It does this secure in its power to shape the rules of the market and in its ability to use trade-distorting instruments (in sectors where it is less strong). Thus the EU can be understood as adopting an instrumental or selective (moderated) neo-liberal approach to external economic policy, and this is reflected in its approach to regionalism.

Regarding the forms of institutional co-operation (the institutional forms of regionalism) that the EU promotes, Bicchi (2006) has explained how the EU tries to project its own models outwards in an 'our size fits all' policy. This begs the question as to what the EU's model precisely is, which has been much disputed. Given the apparent functionalist ethos of the UfM it is worth considering the original debates on European integration and functionalist theory. Given its incremental, economistic, focus, the EU/EEC was clearly related to the functionalist ideal. The core academic theory that sought to explain it was neo-functionalism which focused on the functional spillover into new areas of co-operation, the role of Brussels institutions that attempt to further integration and the coalescing or Europeanization of interest groups (Haas, 1958). Yet the originator of functionalist theory, Mitrany (1965), was highly critical of the EEC. To understand why, it is worth reviewing the core tenets of functionalism.

Writing in 1944, Mitrany argued strongly that that ambitious, legalistic, international or continental unions would not be a sustainable path to peace and co-operation (Mitrany, 1944: 10–19). Yet the solution was not the free market either.

Mitrany was a socialist, inspired by interventionist projects such as the Tennessee River Valley authority in America. His vision of an integrated world was not based just on private businesses and trade, but on the role of technocrats engaged in practical co-ordination and international public agencies in specific sectors. Core tenets of his proposal were that the membership, decision-making and form of such agencies would be entirely flexible (elastic) depending on the function involved. In avoiding political disputes and organizational rigidities, these would be able to better support development of human welfare and help overcome atavistic nationalist tendencies. At the same time they would not be overtly threatening as they would be limited to practical sectors (although he hoped that the agencies would gain more political power and legitimacy in time). It was all about developing more diffuse forms of power and authority in different sectors where transnational co-operation could produce real benefits.

In contrast, neo-functionalist theory described (and approved) a form of centralization of power; the very definition of integration was that a 'new centre of power' was created in Europe, to which political and economic interest groups would gravitate (Haas, 1958). European integration was politicized and based on identity and territory. Mitrany decried the implicit (and sometimes explicit) federal tendencies of neo-functionalist theorists and practitioners (in the Commission). He feared that the expansive, comprehensive, 'self-inflating' EEC had the potential to become a regional power bloc and develop a form of macro-nationalism (Mitrany, 1965: 142). Thus, while European integration was aligned with functionalist ideas in some respects, the 'neo' in neo-functionalism heralds quite a significant difference. This state-like, power-centric tendency of the EU has also been reflected in its external relations, based on the gravitational pull of its internal market. It attempts to legalize and strengthen the economic interdependence between it and its neighbours and then to create joint institutions; or to advise on the formation of separate regional institutions. Thus the EU has embodied neo-functionalist dynamics within and without. Furthermore, it has tended towards a neo-liberal approach to promoting regionalism/regionalization. Let us turn to the Mediterranean to further illustrate this.

## The EU's Policy Acquis in the Mediterranean

While the EU/EC has been engaged in substantial economic diplomacy in the Mediterranean since the Global Mediterranean Policy of the 1970s, it was only in the post-cold war period that it began to develop serious policies for regionalism and regionalization. The Euro-Mediterranean Partnership (EMP), launched in 1995, established the objective of a Euro-Mediterranean Free Trade Area by 2010, thus inextricably linking regionalism in the Mediterranean to the EU. In reality, the major form of legal integration pursued was bilateral, with individual EU–Mediterranean partner country agreements (the association agreements) that provided for staged liberalization of nearly all industrial products. The major regional dimension of EU policy here was focused on a (not very smooth) process of harmonizing rules of origin which should serve to further intra-Mediterranean trade. The association agreements only fastened down free trade in goods but envisaged much deeper free trade and regulatory harmonization. Apart from legal integration agreements, the EMP developed numerous formal institutions at the regional level. Eight Euro-Mediterranean ministerial meetings have been held from 1995 to 2010. They are the primary institutional framework promoting free trade. Other institutions deal with socio-economic co-operation separate from the trade liberalization agenda. Regular ministerial meetings and co-operative policies have been launched in key sectors such as the environment, energy and transport. For example, the energy partnership involves technical and financial co-operation and well as a push for market integration.

The European Neighbourhood Policy was developed (in 2003) mainly with eastern Europe in mind but was also expanded to the Mediterranean, where it sat uneasily with the pre-existing policy framework. As the major policy objective of the ENP was bilateral integration with the European market, this seemed to contradict the regionalism of the EMP. Or, at any rate, the EU's commitment to regionalism was even more EU-centric: the presumption being that harmonization of laws with the EU's internal market would lead, amongst other things, to South–South integration.[2]

However, it was clear that only a few Mediterranean partners, such as Israel, Morocco and Tunisia, were willing to commit to this level of integration. Thus arguably EU policy was fragmenting rather than integrating the economies of the Mediterranean. It had also fragmented the Arab Maghreb Union from the start of the EMP as Mauritania and Libya were not included. On the other hand, the pressure of EU trade liberalization policies, combined with global pressures, has encouraged other intra-Mediterranean trade arrangements most notably the Agadir Agreement (2004) which has developed free trade between Egypt, Morocco, Jordan and Tunisia. This is a lightly institutionalized regime (with a small secretariat based in Amman). There have also been numerous bilateral intra-Med free trade agreements, many involving Turkey (which ironically has had little interest in the EMP institutions as it focused on accession).

The EMP has continued to develop, in tandem with the ENP and there have been innovations in the form of the Euro-Med Parliamentary assembly, which is another mode of formal regionalism. It is self-evident that the original goal of a fully fledged free trade area by 2010 has not been realized but the process continues. Since 2003 the Commission has been harmonizing the rules of origin (developing cumulation of origin) to facilitate transregional production networks. In December 2009 the Euro-Med trade ministerial meeting agreed a road map for further trade liberalization (including reducing non-tariff barriers) and an expansion of the cumulation of origin system to include the Balkan countries. Commissioner Ferrero-Waldner announced that 'this is the Union for the Mediterranean at work' (ENPI Info Centre, 2009), but this was clearly a process that long pre-dated the Sarkozy initiative.

Apart from multilateral diplomacy and institutionalization, aid policy was the major EU tool to develop regionalism. The Mesures d' accompagnement (MEDA) aid programme had regional integration as a specific aim and a proportion of the budget was devoted to this. Regional Strategy Papers were developed to complement the bilateral aid. Much of this was devoted to further networking among policy makers in an effort to socialize support for reform and integration. For example the Euro-Med Market Programme involved technical studies of legal/regulatory barriers to trade and then training of officials to try and minimize the obstacles (Sánchez Monjo, 2005). MEDA also contributed technical and financial support to the Agadir agreement, focused on boosting the capability of the Secretariat to develop the agreement. The Commission has developed a sort of speciality in supporting regional institutions (note its active role with Mercosur, the Association of South East Asian Nations (ASEAN) and the sub-Saharan African institutions) although it is clear that aid on its own can have a limited impact. While the ENP was explicitly bilateral, the aid instrument created to support it (which replaced MEDA in 2007) was more concerned with developing transnational co-operation at numerous institutional and geographical levels. This European Neighbourhood Partnership Instrument has a rather confusing range of regional and cross-border strategy papers (as well as bilateral papers and a 'neighbourhood'-wide paper). The 2007–10 regional programme allocated funds to support the integration of energy markets (harmonization of regulations), reform of transport policy as well as general regulatory harmonization and reforms following on from the Euro-Med Market programme.

The policies supporting regionalism described above are, obviously, also designed to foster regionalization. There is also direct aid for regionalization (in fact this preceded the formal regionalist efforts of the EMP). As a part of the EU's Renovated Mediterranean Policy (1991) the Commission launched decentralized co-operation programmes geared to building contacts between local urban governments, academic institutions and media organizations throughout the Mediterranean. Alas these had to be disbanded and reformed due to problems with the financial arrangements in question (Giammusso, 1999: 26), one example of the prosaic but serious challenges in developing transnational aid projects. The MEDA regional envelope supported a range of activities beyond public institutions. It funded the development of networks of economists (Forum Euroméditerranéen des Instituts de Science Économiques, Femise) and security experts (Euro-Mediterranean Studies Commission, Euromesco). Other programmes funded include the Euro-Mediterranean Information Society (EUMEDIS) to facilitate exploitation of the internet and new information technologies of the countries in the region and EU Med Connect to connect up the national research networks of most of the Mediterranean partners (to facilitate co-operative applications and joint projects). EU aid has also consistently encouraged networking between other elements of society, with a particular focus on youth and cultural dialogue. The Anna Lindh Foundation for intercultural dialogue, which was inaugurated in Alexandria in 2005, is one example of a tangible joint project.

The European Neighbourhood and Partnership Instruments (ENPI) regional programme for 2007–10 had a major focus on developing the actual transport, energy and telecommunications infrastructure of the region and direct efforts to support flows of foreign direct investment. This decade also saw various efforts to access more hard finance for Euro-Med relations. The Neighbourhood Investment Facility uses grants from the Commission and European governments to leverage concessional loans from international financial institutions. The European

Investment Bank, an autonomous EU institution which had been active in the Mediterranean for some time, established the Facility for Euro-Mediterranean Investment and Partnership (FEMIP) in 2002 to facilitate funding for the private sector in the region. It should be noted before concluding this section that EU co-operation and aid for different forms of regionalism and regionalization has synergies with the US, non-governmental organizations and international economic institutions support for similar objectives (with a 'Middle East' rather than a Mediterranean framework).

Clearly, the EU's Mediterranean policy is hard to neatly characterize. It has oscillated between a truly regionalist (in the pan-Mediterranean sense) and a more bilateral approach. Its economic integration policies are not as explicitly neo-liberal as the North American Free Trade Area. Regional co-operation has not just focused on trade liberalization (but on co-ordinating common policies in key sectors). Aid is an intrinsic part of the relationship, and while much of this is devoted to free market reforms (as outlined above) a substantial proportion goes to support socio-economic cohesion (Holden, 2009: 59).[3] The regionalization it seeks to engender goes beyond that of market forces. However, notwithstanding these caveats it is apparent that the primary thrust of the EU's economic policy is based on 'negative integration'[4] into the EU internal market. Also, the overall amount of financial support the EU offers is, given the scale of the region and the impact of trade liberalization, not large. In that sense, the EU's Mediterranean policy has been essentially neo-liberal in operation.

## Enter the UfM

The state of play in the Mediterranean in 2007 did not give great cause for optimism. Regarding the regionalist vision, there was still acute disunity, with conflicts and tensions in the Levant, and in the Maghreb, constricting intra-Mediterranean and Euro-Mediterranean integration. The non-attendance of the leadership of the Mediterranean partners at the tenth anniversary summit was a dramatic illustration of a lack of interest in EU-led regionalism at the very highest level (although it should be noted that that the EMP multilateral level had never gone beyond the level of foreign ministers). The EU had succeeded in agreeing association agreements with nearly all of the Mediterranean partners, thus laying the basis for a form of economic regionalism based on its own market). While there have been many powerful criticisms of the social effects of EU trade policies, they have led to significant increase in trade and investment between the EU and North Africa (Eurostat, 2007: 3–4) and to a degree of regionalism and regionalization between sub-sets of the Mediterranean partners. Thus, although the results of its policies had not been very impressive, to regard the EU's Mediterranean policy as an outright failure is slightly extreme. However regionalization (even in the economic sphere) has been relatively disjointed and fragmented, as it is centred on European interaction with specific countries (Eurostat, 2007: 3–4). More comprehensive regionalization (apart from migratory networks, the one topic that the EU is always less enthusiastic about) has been lacking.

Within this context President Sarkozy launched his idea of a Mediterranean Union in 2007. The birth of the UfM is very much related to high politics. In this case the context was the French presidential election of 2007 and Sarkozy's desire to offer a foreign policy vision, and an alternative to Turkish entry into the EU itself. After considerable intra-EU squabbling the UfM was established in 2008. (At the core of the arguments was the desire of other member states, especially Germany, and the Commission, to ensure that all EU member states were included and that the new institution was harmonized with the existing EU policy framework.) The broad principles were agreed at the Paris summit of heads of state. This high profile event improved the visibility of Euro-Mediterranean relations substantially, although few mainstream media reporters evinced any knowledge of the previous EU policy acquis in the region.

Certainly, the ethos of the proposed new Union was a departure from the bilateral focus of the European Neighbourhood Policy (which was in itself a departure from the EMP's regionalist discourse). Also the inclusion of new countries in the Balkans and North Africa (Mauritania, with options for Libya to join also) gave it a more comprehensive Mediterranean dimension. To the East of the EU, over a roughly similar timeframe, more traditional geographical regionalism also reasserted itself against the ENP template. On the initiative of the Polish and Swedish governments a new Eastern Partnership (EaP) was launched by the EU in March 2009. This was to be applied to the near Eastern neighbours Armenia, Azerbaijan, Belarus, Georgia, Moldova and Ukraine. However, the EaP was less of a departure from the EU's traditional *modus operandi* than the Union for the Mediterranean. Its major instruments remain trade liberalization and increased Commission-managed aid. Although less 'original', it would seem to offer more in terms of actual integration than the UfM (Martin, 2009: 243).

A summit of foreign ministers in Marseilles in November of the same year helped clarify the institutional framework for the new Union. These institutions represent the obvious innovative dimension to the UfM. There is a co-Presidency (consisting of one EU government and one from the other Mediterranean countries) which is assisted by a joint permanent committee. Whereas regular EMP institutions only went as far as ministerial meetings, the UfM will have a biennial summit of leaders. At this level UfM projects will have to be approved. All of this is designed to help ensure a higher level of partnership at the highest political level (Aliboni and Ammor, 2009: 9). The summit still left many things unclarified however, including how the co-Presidency and the Secretariat would work. These new bodies obviously represent, theoretically at least, a substantial degree of institutionalization of Mediterranean regionalism, and this will doubtless be discussed in other contributions. Other institutions, such as the meetings of senior officials to prepare for ministerial meetings, will be more familiar to observers of Euro-Mediterranean relationships. Most significant with regard to functionalist ideas is the establishment of a 'project-orientated' joint Secretariat, based in Barcelona, to handle the technical aspects of the relationship.

As to the actual substance of the new Union, it is aligned with the broad and holistic objectives of previous Euro-Med agreements (peace, prosperity etc.). The major substantive activity of the new Union is in the field of development and environmental policy. Six priority areas for projects were outlined:

- de-pollution of the Mediterranean;
- maritime and land highways;
- civil protection;
- alternative energies and Mediterranean solar plan;
- higher education and research; and
- supporting business (small to medium sized enterprises in particular).

Obviously these priorities are not in themselves a new departure for the EU, or other international organizations, in the region. What is noteworthy is a greater emphasis in the discourse on visible, tangible development projects that directly affect the lives of ordinary people. This is a departure from, necessarily, elitist legal and regulatory agreements and (to an extent) from EU aid policy, which has focused more on technical assistance and aid for reform than direct development projects. In contrast, the UfM priorities are supposed to be directly geared to concrete activities that will provide jobs and infrastructure, support new industries and perceptively improve the lives of the population. The Commission states that the UfM should support projects that 'ensure reinforced cooperation both at regional and sub-regional levels, have a real potential for regional integration, and are both inclusive and non discriminatory' (European Commission, 2008: 10). These projects can have a variable geometry, they must be transnational in nature but do not need to concern the whole region. The process of project selection is open to initiatives from civil society and business as well as the states concerned. This theoretical process would be in marked contrast to traditional EU aid, in which programming is tightly controlled from Brussels. The general process suggested that the Secretariat will propose projects (after consultation) but then the high officials and the foreign ministers have to collectively approve a project before it has the UfM stamp.

The major question here, as always, is where the funding would come from. It became clear in the run-up to the launch summit that no new funds would be available from the EU's own budget. The hope is for greater funds from the participating states, other international financial institutions and the private sector, based on the good publicity and political commitment offered by the UfM. This is an example of the EU and its partners trying to use their political capital to mobilize actual capital. Creative public–private partnerships seem to be on the agenda. The follow-up meeting in Marseilles also expanded the UfM agenda to encompass many other EMP issues (including free trade, intercultural dialogue). The institutions are still evolving and several proposals have been made for a more institutionalized civil society role, for example a Euro-Mediterranean Economic and Social Committee. The Secretariat is not an implementation body in the literal sense of administering projects, it will serve as an overall co-ordinator and each project will be managed by the lead organization in question.

In brief, the UfM ethos does, to a degree, accord with functionalist principles in its approach to development and regionalization. The development element is, in principle, strongly focused on facing up to practical challenges to the welfare of the citizens of the Mediterranean. The degree of flexibility in terms of project management, funding and participation are very much in line with a functionalist ethos.

Also, implicit in the broader strategy is the hope that successful co-operation in this sphere will lead to a popular acceptance of the institutions (this was Mitrany's hope with regard to his transnational agencies). However, in important respects the UfM departs from the functionalist vision of Mitrany. It is an overarching framework which deals with various socio-economic sectors rather than a purely specialist organization. More importantly, it is highly politicized (the perils of which would soon become evident). Thus the institutional form does not 'follow function'; it has been determined by macro-political diplomacy rather than practical design. High-profile political summits and rhetoric of Union are anathema to 'pure' functionalism.

The UfM thus appears as a rather ad hoc exercise in many respects. It is highly malleable and has the ability to appear to be all things to all people. One UNEP (United Nations Environmental Programme) official expressed the hope that 'energy and environmental security concerns will contribute to enhanced co-operation by offering the two functional pillars on which the Union for the Mediterranean will be able to prove its legitimacy and efficiency', a vision analogous to that of the European Coal and Steel Community (Piet, 2009: 1). This is an interesting and productive vision, but in reality the UfM is much less rigorously focused than the ECSC (which had clear objectives in the sectors in question, and strong institutions). Any kind of comparison is indeed far-fetched and this is an example of how the UfM has inspired many hopes that are frankly unrealistic. On the other hand there has, understandably, been much scepticism from all shores of the Mediterranean as to the point of the UfM. As one Egyptian source put it:

> in our view, UfM is just another layer of changes in the EU policy which just adds vagueness and complexity to EU–SMC relationships ... The advantage of the new institutional set-up could have been easily included in the existing functioning set up of the Barcelona process (through enforcing the existing provisions by setting joint programmes for implementation, and specific deadlines), without a new agreement that remains vague in terms of objectives, means of implementation and funding. (Ghoneim, 2009: 95)

This criticism chimes with the author's own sentiments. Informed observers have recounted a more positive reaction to the core idea of implementing transnational projects on the part of North African and Israeli elites (Aliboni et al., 2008: 23, 26). Indeed, it is hard to see who would not favour the anodyne apolitical UfM projects, although the fact that some partners insisted, in the Marseilles declaration, on articulating their right to oppose a transnational project (paragraph 9) implies that some consider that these projects may have broader socio-political significance.[5] The prospect of a greater intergovernmental partnership is greeted with understandable scepticism: 'a co-Presidency un-backed by any power of decision or conduct would be useless, senseless and would affect no influence over the contents of the Partnership' (Driss in Aliboni et al., 2008: 23).

As to what has actually been done to support development and regionalization, it is far too early to even hint at any kind of evaluation. The Gaza conflict of December 2008 and January 2009 severely disrupted the high politics dimension of co-operation, but the joint permanent committee meetings continued. The establishment of the Secretariat was delayed (primarily due to the political friction related to the Gaza incursion and poor relations between the new Israeli government and many Arab states). Thus we do not have examples of how it will lead core UfM projects. The statutes were finally agreed in early 2010 (apart from some minor squabbles as to the number of deputy directors-general). It is apparent that this Secretariat, with a staff of only 20–25 people, will be far too small to have the option to 'manage' projects in the sense an aid agency does. It will concentrate on a co-ordinating or steering role (and in raising funds in the first place). It is receiving financial support from the European Commission (initially €3 million for 2010). At the inauguration of the Secretariat European Commissioner Štefan Füle (2010) stated that 'We will stand behind the Secretariat as it keeps growing and developing its tasks. And I would be happy to come here again as our work progresses to commit more of our funds and our people to the daunting endeavour that we are starting today.'

Since the Paris summit the Commission has also allocated €90 million from the ENPI budget line on projects related to the priorities identified (this is not 'new' money; it was allocated for the neighbourhood region for the current period). This includes €22 million for an environmental reporting system, €9 million for an 'Invest in Med' initiative and various transport projects. As well as this, there are numerous ongoing ENPI projects which are roughly in harmony with the ethos and objectives of UfM. Apart from the ENPI regional aid discussed in the previous section there are also bilateral cross-border co-operation programmes highly relevant to UfM themes. For example the Italy–Tunisia programme is allocated €25 million to support cross-border activities for sustainable development and general networking. (Again it must be stressed that these should not be understood as a UfM contribution.) The European Investment Bank (EIB)'s FEMIP facility is seen as potentially central to the basic mobilizing aims of the UfM. The European Parliament has joined the call (also made by various Mediterranean partners at different points in time) to establish a full Mediterranean Investment Bank but this has been resisted by the EU Council of Economic and Finance ministers thus far. The EIB and some national European institutions have agreed to help finance up to €5 billion worth of projects related to renewable energy in the region (the Mediterranean Solar Plan). A Solar-Med conference in Paris in 2010 is led by the private sector with the support of public institutions such as the UfM itself and the Commission. The perils of 'high politics' resurged in April 2010 when a ministerial conference on a Mediterranean strategy broke down over a disagreement on how to name the occupied Palestinian territories.

The complications created by the high politics agenda of the UfM bear out Mitrany's original critique of bombastic politicized integration/co-operation schemes. As to what the new institution contributes to mobilizing resources and ideas for regional development is not so clear. The unequivocal focus on promoting development and welfare (as opposed to reform, however desirable) is justified by necessity and may prove useful. Substantial resources have been raised for the flagship Mediterranean Solar Plan, but in reality this could have been done anyway, and it is not clear what transnational dynamic there will be to this. The type of UfM-directed project visualized at the Paris summit has not yet appeared and so cannot be evaluated. It is not clear how the UfM can succeed in ways other efforts have not. The limitations on top-down strategies and aid projects to develop/shape societal and economic forces are easily apparent. In fact one of the reasons international aid policy developed in its current form – of programmes rather than discrete projects – is that project aid without a coherent strategy was judged to be inefficient. Acute barriers to regionalization in the Mediterranean remain, most notably, EU desire to restrict and control migration, and authoritarian policies in most of the partners. Also, there is little reason to hope that the UfM's rhetoric of openness and participation will mean more than any other international/intergovernmental regime.

## Conclusion

The UfM framework is contradictory as it suggests a new, more flexible, functionalist approach to supporting regionalization and development but it also embodies all the complications that high-level political regionalism implies. It is a kind of smorgasbord of policies and institutions and it remains to be seen what its primary form will be. For the moment it serves a political purpose in being all things to all people. With regard to the central question of this paper, it is crucial to emphasize the simple point that the UfM does not replace the pre-existing EU policy frameworks. The ENP, however uninspiring, remains an important channel for governments seeking integration into the EU and access to EU programmes. Countries such as Morocco are pursuing their own advanced status, to a large extent focused on integration into the internal market. (Tunisia and Israel are also more interested in deeper bilateral economic relations with the EU.) Regarding EU policy, its drive for free trade in the region (with all the costs and benefits that entails) continues and the EMP trade agenda has further developed. All of which means that, for better or worse, Euro-Mediterranean relations have not changed very much. The EU's Mediterranean policy has been essentially neo-liberal in ethos and impact (granted that it has included more interventionist and functionalist mechanisms). The UfM ethos may not be in itself 'neo-liberal' but it does not replace the previous policies. It is more of an accretion than a change of direction.

The EU's efforts over the past two decades are open to criticism and their impact, particularly on regionalization, has been disappointing.[6] However, they do represent a substantial effort to use the Union's legal, institutional and aid instruments to shape the region. Given the extent of the political, economic and cultural fragmentation of the Mediterranean, the EU's task was always daunting. The disappointing progress of regionalization is not just because of geopolitical fissures but due to the constraints created by the different political and economic contexts within the partner states. This was very clearly explained by Giammusso in his study of the constraints on effective decentralized co-operation in the 1990s, and his points still apply today (Giammusso, 1999). These constraining factors deserve to be fully considered at the highest political level, rather than being constantly swept under the table by new initiatives and institutions. We will have to wait to study how the Secretariat works to judge whether it can help take regionalization and co-development in the Mediterranean to another level. However, the conclusion has to be that there is no indication that the UfM offers any new ideas to overcome the constraints which have hampered Euro-Mediterranean co-operation from the outset.

## Notes

[1] The policies of the major Western governments in reaction to the financial crisis and economic recession of 2008 are a clear example of this. Many predict/hope that this crisis signals the end of the dominance of global neo-liberalism but this is far beyond the scope of this study.

[2] In other words it could be stated that the ENP implied a more explicit hub and spokes relationship between the EU and its partners. (The EMP had more of a focus on South–South relations). However it is apparent that any free trade arrangement between the EU and its neighbours will be of the 'hub and spokes' variety as the EU is by far the largest market and pre-existing trade patterns already take this form.

[3] Over the period 2002–06, I calculate that approximately 50 per cent of MEDA national funding was allocated to reform-centred activities, 40 per cent went to socio-economic development and the remainder went to support for the private sector or miscellaneous themes (Holden, 2009).

[4] This term refers to a form of integration based on reducing regulatory power at the national level without recreating it at the regional level.

[5] Or perhaps this can best be understood as an authoritarian reflex.

[6] In January 2011 a vibrant trans-regional protest dynamic emerged in the Arab world. This was entirely independent of the various Euro-Mediterranean policies and institutions.

## References

Aliboni, R. & Ammor, F. (2009) Under the shadow of 'Barcelona': from the EMP to the Union for the Mediterranean, EuroMesco Paper 79.

Aliboni, R., Driss, A., Schumacher, T. & Tovias, A. (2008) Putting the Mediterranean Union in perspective, p. 68, EuroMesco Paper 68.

Bicchi, F. (2006) Our size fits all': normative power Europe and the Mediterranean, *Journal of European Public Policy*, 13(2), pp. 286–303.

Breslin, S. & Higgott, R. (2000) Studying regions: learning from the old constructing the new, *New Political Economy*, 5(3), pp. 333–352.

ENPI Info Centre (2009) Focus on free trade area as Euromed ministers meet 08-12-2009, Available at http://www.enpi-info.eu/mainmed.php?id=20272&id_type=1 (accessed 12 January 2010).

European Commission (2008) Barcelona Process: Union for the Mediterranean Brussels, 20/05/08 COM(2008) 319 (Final).

Eurostat (2007) The European Union and its ten Mediterranean partner countries: growing trading links, Available at epp.eurostat.ec.europa.eu/cache/ITY_OFFPUB/_/KS-SF-07-070-EN.PDF (accessed 17 December 2009).

Friedman, M. (2002) *Capitalism and Freedom* (Chicago: University of Chicago Press).

Füle, S. (2010) Speech by the European Commissioner for Enlargement and Neighbourhood Policy at the inauguration of the seat of the Secretariat of the Union for the Mediterranean, 4 March, Available at http://europa.eu/rapid/pressReleasesAction.do?reference=SPEECH/10/67&format=HTML&aged=0&language=EN&guiLanguage=en (accessed 14 April 2010).

Giammusso, M. (1999) Decentralised cooperation networks: civil society initiatives and the prospects of economic development, *Mediterranean Politics*, 4(1), pp. 25–62.

Ghoneim, A. (2009) Does Egypt need this change in EU trade policy, in: M. Comelli, A. Eralp & C. Ustun (Eds) *The European Neighbourhood Policy and the Southern Mediterranean* (Ankara: Middle East Technical University Press).

Haas, E. B. (1958) *The Uniting of Europe: Political, Social and Economic Forces 1950–57* (Stanford, CA: Stanford University Press).
Hayek, F. (1979) *Law Legislations and Liberty, Vol. 3: The Political Order of a Free People* (Chicago: University of Chicago Press).
Holden, P. (2009) *In Search of Structural Power: EU Aid Policy as a Global Political Instrument* (Aldershot: Ashgate).
Krugman, P. (1986) *Strategic Trade Policy and the New International Economics* (Boston: MIT Press).
Martin, I. (2009) EU–Morocco relations: how advanced is the advanced status? *Mediterranean Politics*, 14(2), pp. 239–245.
Mitrany, D. (1944) *A Working Peace System* (London: RIIA).
Mitrany, D. (1965) The prospect of integration: federal or functional, *Journal of Common Market Studies*, 4(2), pp. 119–149.
Piet, R. (2009) Energy and environment: the new coal and steel of the Union for the Mediterranean, Paper for the European Union Studies Association conference, Los Angeles, 23–25 April
Polanyi, K. (2001) *The Great Transformation: The Political and Economic Origins of Our Time* (Boston, MA: Beacon Press).
Sánchez Monjo, E. (2005) *Regional Programme for the Promotion of the Instruments and Mechanisms of the Euro-Mediterranean Market 2nd Phase* (Maastricht: European Institute for Public Administration).
Tovias, A. (2000) Trading blocs and international relations: economic groupings or political hegemons? in: T. Lawton, J. Rosenau & A. Verdun (Eds) *Strange Power: Shaping the Parameters of International Relations and International Political Economy* (Aldershot: Ashgate).
Van Der Pijl, K. (2006) A Lockean Europe, *New Left Review*, 37 (January–February), pp. 9–37.

## Bibliography: Other Sources Consulted

Balfour, R. & Schmid, D. (2008) Union for the Mediterranean: disunion for the EU?, EPC Policy Brief, February.
Cvikl, M. (2008) Report of the Parliamentary Assembly of the Mediterranean, 18 November Report on free trade in the Mediterranean – strengths, weaknesses, and future development.
Euro-Mediterranean Foreign Ministers, Final Statement Marseilles, 3–4 November 2008.
European Commission (1997) Supplementary information and analysis of the results of the evaluations of the decentralized cooperation programmes in the Mediterranean.
European Commission (2007) ENPI regional strategy paper for the Euro-Mediterranean Partnership 2007–13.
Hunt, D. (1999) Development economics, in: G. Joffe (Ed.) *Perspectives on Development – The Euro-Mediterranean Partnership* (London: Frank Cass), pp. 16–39.
Martin, I. (2004) The labyrinth of subregional integration in the South Mediterranean, *Med. 2003. Mediterranean Yearbook*, pp. 165–168 (Barcelona: Fundación CIDOB/Instituto Europeo del Mediterráneo).
Rosamond, B. (2000) *Theories of European Integration* (Basingstoke: Macmillan).

# The UfM and Development Prospects in the Mediterranean: Making a Real Difference?

DIANA HUNT
Independent Researcher and Consultant, UK

ABSTRACT  *Can the UfM make a distinctive contribution to development in the Mediterranean region? How innovative are its priority project themes? In an attempt to answer these questions, this contribution examines the relationship between the priority project themes of the Union for the Mediterranean and the substance of the Euro-Mediterranean Partnership (EMP) and the European Neighbourhood Policy (ENP) which preceded it. It explores the extent to which, by establishing the six priority project themes, the UfM has broken with the previous focus of the economic reform programmes and projects promoted by the EMP and the ENP. It then reviews the potential contribution of the themes to economic development and examines selected factors which may influence project implementation and goal attainment, including funding access and the impact of the UfM's institutional innovations on the effective pursuit of project themes. The conclusion is that there is scope for the UfM to enhance development in the Mediterranean region but outcomes so far have been limited and future outcomes depend on a range of variables including geo-political tensions, national priorities of partner states and funding availability.*

This contribution explores the extent to which the six priority project themes of the Union for the Mediterranean (UfM) represent a break with, or continuation of, the orientation of the Euro-Mediterranean Partnership (EMP) and the European Neighbourhood Policy (ENP) and the potential development impacts of these themes. The paper starts with a brief account of the core approach of the EMP in the economic sphere and of the varied response to it. In what follows, development itself is given a primarily economic-cum-environmental interpretation as a process of economic growth combined with reduction of unemployment, poverty and extreme asset inequality and with protection of non-renewable resources. This interpretation focuses on outcomes but it also entails, if development is to be self-sustaining, the transformation of productive structures and the development of an endogeneous technological capability and an appropriate organizational capability underpinned by institutions which are effectively enforced – *i.e.* development as *process* as interpreted by the 'new developmentalists' (see Khan, 2010: 6–7).

A number of themes recur. They include the significance of the EU's role as driver both of the EMP and ENP and of the institutional structure and focus of the UfM, the diversity of economic structure and policy priorities of partner states, the continuing importance of institutional reform for the Euro-Med enterprise and the potential impact of the UfM's own institutional innovations on the effective pursuit of project themes.

Table 1 highlights the variation between the economies of seven of the original Mediterranean partner economies in the Barcelona Process (BP), plus Libya, in size, per capita income and sectoral composition of GDP. In Algeria and Libya over 90 per cent of the contribution of industry to GDP comes from energy and mining and less than 10 per cent from manufacturing. The major fuel exporters are Algeria and Libya plus Egypt; Syria is still just a net exporter and Tunisia a net importer following depletion of known oil reserves: Morocco, Jordan and Lebanon import almost all their fuel. Morocco, Egypt and Syria have large shares of agriculture in GDP, and still larger shares of the labour force employed in this sector, while the GDP shares of agriculture in Algeria and Tunisia are also enough for large variations in agricultural output to affect GDP growth rates. The economies bordering the eastern and southern Mediterranean have a Mediterranean climate along their coastal strip with semi-arid and arid hinterlands. They are exposed to volatile rainfall whose impact on farm output and GDP is exacerbated by the low proportion of irrigated farm land (Tables 1 and 2). In those economies heavily oriented to oil production variations in world oil prices also affect not only export revenues but the value of total GDP. High oil sector shares in exports and GDP have tended to be associated with high foreign exchange reserve to GDP ratios over relatively long periods, giving some of these economies greater scope than others to resist external pressures, including aid conditionality, to liberalize their foreign trade and internal markets.[1]

While differences in economic structure influence the growth potential and growth performance of these economies, the latter also differ both in political conformation and in government attitude towards the economic reforms which have been promoted by the EU, first under the EMP and then under the ENP, and in policy stance to towards the key issues which dominate the southern and eastern Mediterranean region. Political, rather than economic, variables held back Syria's full participation in the economic chapter of the EMP, as well as Libya's involvement.

**The Goals and Development Significance of the EMP and ENP**

The 1995 Barcelona Declaration was 'intended to establish a comprehensive Euro-Mediterranean partnership in order to turn the Mediterranean into a common area of peace, stability and prosperity through the reinforcement of political dialogue and security, an economic and financial partnership and a social, cultural and human partnership' (Europa, 2010).

The main driver of the EMP and the ENP has consistently been the EU. For the dominant partner, development in Mediterranean partner states[2] was an instrumental

**Table 1.** Key statistics for eight economies of the southern and eastern Mediterranean, 2008

|  | Algeria | Egypt | Libya | Morocco | Syria | Tunisia | Lebanon | Jordan |
|---|---|---|---|---|---|---|---|---|
| GDP (current US$) (billions) | 173.9 | 162.8 | 99.9 | 86.3 | 55.2 | 40.2 | 28.7 | 5.1*** |
| Population, total (millions) | 34.4 | 81.5 | 6.3 | 31.2 | 21.2 | 10.3 | 4.1 | 5.9 |
| GNI per capita, (current US$) | 4,260 | 1,800 | 11,590 | 2,580 | 2,090 | 3,290 | 6,350 | 3,470 |
| Agriculture share in GDP (%) | 9 | 14 | 2 | 16 | 20 | 10 | 5 | 3 |
| Irrigated arable land (%)* | 2 | – | – | 5 | 8 | 4 | 10 | 8 |
| Industry share in GDP (%) | 69 | 36 | 78 | 20 | 35 | 28 | 22 | 34 |
| Manufacturing share in GDP (%) | 5 | 16 | 4 | 14 | 13 | 18 | 10 | 20 |
| Services share in GDP (%) | 23 | 50 | – | 64 | 45 | 62 | 73 | 63 |
| School enrolment, primary (%) | 95.4* | 95.7* | – | 88.8* | 94.5** | 95* | 83.2 | 89* |
| Population growth (annual %) | 1.5 | 1.8 | 1.9 | 1.2 | 3.5 | 1.0 | 1.0 | 3.2 |

*Notes:* *Relates to 2007; **relates to 2001 (debt) or 2002 (enrolment).
EIU reports 8% irrigated land for Tunisia.
*Source:* World Bank, Key Statistics (2010), except ***Economist Intelligence Unit Country Profiles (2008).

**Table 2.** Rainfall quality and agriculture and GDP growth rates, Morocco

| Year | Rainfall* | Agric. Growth (%)** | GDP growth (%)** |
|---|---|---|---|
| 1995 | Extremely poor | −44.0 | −6.6 |
| 1996 | Very heavy | 78.0 | 12.2 |
| 1997 | Poor | −27.0 | −2.2 |
| 1998 | Average to good | 25.0 | 7.7 |
| 1999 | Poor | −16.7 | −0.1 |
| 2000 | Poor | −15.7 | 1.0 |
| 2001 | Good | 27.6 | 6.3 |
| 2002 | Average to good | 5.6 | 3.2 |
| 2003 | Good | 18.8 | 5.2 |
| 2004 | Below average | 1.9 | 4.6 |

*Notes:* *In previous November–March; **real growth rates.
*Source:* Economist Intelligence Unit (2006: 36).

objective: a non-military means to promote enhanced security within the EU by providing improved employment opportunities and living standards within partner states, thereby reducing incentives for the disaffected to join fundamentalist movements and migrate to Europe. Economic liberalization and modernization in the economies of the southern and eastern Mediterranean also offer the prospect of expanded outlets for FDI. Political and cultural dialogue were each expected to contribute to an increase in mutual understanding and shared values. However, as other contributions to this collection also suggest, it is the economic chapter that was the main focus of the EMP and it is the focus of what follows.

Under the EMP, and then under the ENP, economic integration between the EU and partner states has been promoted through programmes for partner state liberalization of imports of EU industrial goods plus associated liberalization and structural adjustment of their domestic economies, including regulatory reform.[3] Institutional reform in Mediterranean partner states – including domestic market deregulation and privatization of state enterprise – was seen by the EU as key to the success of the liberalization programmes. Together with investments in physical and social infrastructure, these were expected not only to raise efficiency and competitiveness but to create enhanced opportunities for FDI.

The approach adopted by the EU in formulating the substance of the policy reforms which it promoted in partner states represented the application of neo-liberalism to advance desired outcomes (increased economic efficiency within partner states, EU access to southern and eastern Mediterranean markets and enhanced opportunities for FDI)[4] combined with a pragmatic rejection of the neo-liberal approach where this failed to serve EU interests: hence the initial omission of agricultural goods from trade liberalization. The reform programmes were promoted through regular meetings between EU and partner ministers and senior officials focusing on policy dialogue and on capacity building.

From the perspective of those in the partner states who supported the proposed reforms, the latter were seen as essential if their respective economies were to be able to compete in a capitalist global economy. However, not all partner states, nor all interest groups within them who were aware of the proposed reforms, were equally enthusiastic. At government level, this variation in attitude reflected differences in political conformation and imperatives as well as in economic structure between partner states. Some perceived greater advantage in closer ties with the EU than did others. Apart from Turkey, Israel, Cyprus and Malta, where economic links with Europe were already well advanced, Tunisia (which had begun a programme of structural adjustment in 1986) moved fastest to establish a formal association agreement with the EU.

It was intended that the Barcelona Process should promote regional integration both between partner states and between them and the EU. However, in practice the main focus of the BP was on the latter, formally embodied in association agreements between the EU and individual partners. Uptake of the modernization challenge embodied in an association agreement, which included, in its basic format, commitment by the partner signatory to full liberalization of industrial imports from the EU over a 12 year period, was varied, reflecting wide variation in economic and political preparedness and commitment between intended partners. Tunisia signed an Association Agreement first, in 1995, which came into force in 1997 (at the time the Tunisian government and the IMF suggested that implementation of the agreement could add 1–3 per cent to the annual growth rate). Jordan and Morocco signed agreements which came into force in 1999 and 2000 respectively. Algeria and Egypt signed agreements in 2001 which came into force in 2004 and Lebanon signed in 2002. Egypt's negotiations lasted 6 years and when the agreement was finally concluded it diverged from the 12 year standard period for import liberalization, allowing 15 years for removal of tariffs on some goods (luxuries and cars). Although Jordan was an early signatory, the agreement appears to have had little impact: according to one recent report, the EU has financed a series of programmes to help local companies raise their standards and their knowledge of EU markets but progress remains modest. This may be partly because Jordan has a relatively low share of imports from the EU (24.7 per cent in 2007) and sends only 3.5 per cent of its exports to the EU (Economist Intelligence Unit, 2008c: 45). In recent years, Jordan has had a stronger export relationship with the US, also underpinned by a free trade agreement. The late signings by Egypt, Algeria and Lebanon mean that any stimulus to local productive efficiency arising from tariff elimination has not yet been fully felt because tariff reductions on finished goods which compete directly with local products come in the final years of tariff elimination (tariffs on raw materials, capital goods and intermediate goods are eliminated first).

The EU supported reform programmes with financial and technical assistance, initially under the MEDA programme and then the European Neighbourhood and Partnership Initiative (ENPI). The sums allocated by the EU under MEDA for disbursement to partner states were not insignificant: €3.435 billion for 1995–99 (with large shares going to Egypt and Morocco) and €5.35 billion for 2000–2006 (SEMIDE-EMWIS, 2006; Europa, 2009: 1–2). Over the same period the European Investment Bank (EIB) was also mandated to lend over €11 billion to Mediterranean partner states (SEMIDE EMWIS, 2006). For MEDA I, application procedures were cumbersome and disbursement rates low. However, amendments introduced in Phase

II raised the commitment: disbursement ratio to almost 90 per cent (European Parliament, 2007: 5; Morin, 2005: 8). Funds and technical assistance were designated mainly for budgetary and foreign exchange support, plus support for investment in transport infrastructure and water resources and for institutional reform and capacity building. When, in 2007, the Commission introduced the ENPI to replace MEDA, the basis for fund allocation was modified but in principle a country which was advancing political, economic and social reform and market liberalization according to an agreed programme could access at least as much under the ENPI as under MEDA. The EIB continued to act as an important source of credit. Meanwhile, partner countries also received technical and financial assistance from other multilateral sources including the World Bank, the UN and the Arab and African Development Banks, as well as from bilateral sources. The EU was not always the lead source.

The two countries which offer greatest scope for exploring the impact of the economic chapter of the EMP, because they have moved furthest with reform implementation, are Tunisia and Morocco. In both countries the agreements have given a significant push towards deregulation and privatization. While both weaknesses in available data and methodological problems prohibit precise quantitative assessment of macro-economic impacts, it is possible to make some tentative comments on outcomes to date.[5]

Among the partner Arab states, Tunisia was best placed to respond positively to the stimuli provided by a combination of liberalization of industrial imports from the EU (91 per cent of all Tunisia's industrial imports) and the harmonization of regulations and standards. Tunisia had begun a programme of import liberalization and structural adjustment in 1986, and had also begun to implement – albeit slowly – a privatization programme, with IMF and US support. Key aspects of Tunisia's economic structure – the small size of its economy, the decline in its oil sector, the climatic vulnerability of agriculture – plus the volatility of world markets for phosphates and tourism, meant that Tunisia was impelled towards a strategy of diversified export-led growth with a leading role for manufactures. The association agreement with the EU gave new impetus to liberalization, as well as a specific time frame. Relative to Morocco, the manufacturing sector was more diversified. Transport and energy infrastructure were relatively well developed and, despite some notable weaknesses, so too was education. In 1996 Tunisia launched a *mise à niveau* programme, supported by the World Bank, the US, the EU and the African Development Bank, to help firms modernize their management, technology and financial structure. Although uptake was initially below target, the programme is reported ultimately to have achieved significant success. Other aspects of economic reform, including deregulation and privatization, gathered momentum, with the result, inter alia, that Tunisia's telecommunications industry – key to attracting FDI – has recently achieved dramatic growth and increased efficiency. In the clothing sector, FDI has aided Tunisia in confronting the challenges posed by Asian competition in Europe and annulment of the Multi-fibre Agreement (MFA) in 2005.[6]

Between 1996 and 2008, there is no evidence of an upward growth trend in Tunisia (the growth rate averaged 5.6 per cent from 1996 to 2000 and 5.7 per cent from 2006 to 2008, with a dip in the intervening five years partly due to the downward trend in the clothing sector), but the economy has been sufficiently diversified to withstand shocks stemming from occasional poor performance in key sectors, with annual GDP growth in all years except two over this period of between 4 and 6 per cent (World Bank, 2009). Although the predicted increase in Tunisia's growth rate has not yet been forthcoming, arguably it should not have been expected until after liberalization and associated measures were complete (in 2009–10), and both foreign and domestic firms had had time to consolidate their responses. Meanwhile, according to official statistics the unemployment rate fell from a peak of 16 per cent in 1996 to 14.1 per cent in 2007 but these figures are regarded as unreliable. Actual unemployment is thought to be higher and growth so far has not been high enough to significantly dent the absolute numbers of the unemployed (Economist Intelligence Unit, 1996: 15; 2008b: 25).

After 1995, Morocco too moved ahead with a *mise à niveau* programme, albeit more slowly[7], and with infrastructure investment and liberalization of telecommunications, and has subsequently also attracted new FDI.[8] However, differences with Tunisia include lower GDP per head, much higher wealth inequality, a much greater proportion of the labour force engaged in agriculture – in 2008, still at least 40 per cent, unchanged from 1995, compared with 16 per cent in Tunisia, down from 33 per cent in 1995. Morocco's GDP reveals greater volatility than Tunisia's in response to fluctuations in agricultural output. The failure of the policy reforms promoted under the BP to respond to the need for an appropriate agricultural strategy, recognizing a role for intensification of small-scale agriculture which could in turn stimulate the development of output and employment in linked non-farm activities, has entailed a major opportunity cost: not only has growth in rural output and living standards been held back but so have labour absorption and skill formation and the expansion of the domestic market.

Nonetheless, despite some lost opportunities, it would be inaccurate to state that the economic chapter of the EMP, and follow-on programmes under the ENP, have failed to impact on economic performance in either Tunisia or Morocco. In Tunisia, where implementation is most advanced, liberalization of the internal and external sectors, combined with implementation of the *mis à niveau* programme, is reported to have increased the economy's ability to attract export-oriented FDI and to have stimulated some increase in industrial competitiveness (see EIU, 2006: 40, 48; 2008b: 28). In Morocco, regulatory reform and investment in physical infrastructure are also reported to have contributed to raising both domestic business start-ups and FDI (EIU, 2008a: 21–3, 28, 35, 40–41). In other partner Arab states implementation started later and is further from completion.

While the EMP was intended to promote regional development and integration, in practice the focus was predominantly on implementation of bilateral trade liberalization with the EU and domestic national economic reform, both supported by financial and technical assistance from the EU. This outcome reflected both the difficulties which were encountered in promoting trade liberalization between partner states (although some sub-regional progress was made) and the EU's on-going concerns to enhance trade and FDI with its Mediterranean partners. Following the establishment of the UfM, EU balance of payments and budget support for strategic policy reform in partner states has continued under the ENPI. Meanwhile the UfM has highlighted key infrastructural programmes and given increased impetus to some of them: the focus of the MEDA-funded physical and social infrastructure projects (notably transport, energy, water and waste management and education) is echoed in UfM priority project themes. A further element of continuity from the Barcelona Process to the UfM lies in recognition of the need for regulatory reform and harmonization pertaining to some of the priority project themes, in order to enhance efficiency and facilitate new investment.

## The Goals and Priority Project Themes of the UfM and their Relationship to the EMP and ENP

The UfM's founding Summit Declaration represents the Union as a multilateral partnership which will operate alongside the bilateral programmes of the ENPI. The UfM is intended to pursue the goals of the Barcelona Process while enhancing their attainment by addressing three identified weaknesses: the need to upgrade the Partnership's political status, to increase co-ownership among partner states and to raise the partnership's visibility. The Declaration highlighted the UfM's emphasis on institutions and processes already established as part of the Barcelona Process plus certain innovations which are intended to address these three weaknesses: biannual summits, co-Presidency and Secretariat, the latter with technical responsibilities in the fields of project identification, design, funding co-ordination and monitoring (see Johansson-Nogués, this collection). The UfM is to be made more visible by focusing on the implementation of regional and sub-regional projects that will be 'relevant to the citizens of the region' and 'enhance the flow of exchanges among the people of the whole region' (Union for the Mediterranean, 2008: 4). The six priority project themes are:

- de-pollution of the Mediterranean;
- maritime and land highways;
- alternative energies;
- civil protection;
- higher education and research (establishment of a Euro-Mediterranean University); and
- the Mediterranean Business Development Initiative (support for small, medium and microenterprise).

The project orientation of the UfM gives it a narrower economic focus than either the EMP or the ENP. Effectively, the UfM has carved out a sub-section of the former for enhanced emphasis.

While the EU is committed to playing a significant role in assisting funding mobilization from IFIs and private sources for UfM projects, its own funding will come from the ENPI. This entails that UfM projects should be consistent with national action plans agreed under the ENP (so far not, in practice, a significant issue).

In considering the potential development impact of the UfM, key issues to be addressed include the substantive nature of the priority project themes, the scope for mobilizing funding and the potential impacts on implementation of project themes of the measures taken to raise the UfM's political status and to increase co-ownership. The next section focuses on the substance of project themes, on aspects of continuity and change in their selection and the progress made in advancing them. It will be argued that two of the three physical infrastructure project themes contained significant elements of continuity with programmes and projects previously promoted under the BP and the ENP. These themes are discussed first in the following sub-sections, followed by a review of the other themes. Thereafter, the final section provides an overview of the potential contribution of the priority themes to economic development among UfM members, particularly the partner states, and then considers aspects of the UfM's design and funding arrangements which may impact on project theme viability and performance.

*De-pollution of the Mediterranean: The Horizon 2020 Programme*

Protection of the Mediterranean from both maritime and land-based pollution has been a focus of concern at least since the 1970s. The Barcelona Convention, signed in 1976 and for which the United Nations Environment Programme (UNEP) was the lead agency, reflected this concern, committing non-EU Mediterranean cities with populations over 50,000 to install adequate sewage treatment systems by 2010 (2005 for cities of over 100,000). The EU's concern to protect the Mediterranean meant that various waste management projects were instigated in partner states under the MEDA programme. These involved sub-groups of EU and partner states, with a focus on capacity and institution building, including transfer of innovative water management technologies.[9] Meanwhile, the World Bank runs a Mediterranean Environmental Technical Assistance Programme (METAP) dedicated to the design and enforcement of national regulatory frameworks. Horizon 2020 has been described by the EU Commission as designed to 'enhance and catalyse co-ordination' between various actors and to promote new projects where key, bankable gaps are identified (Commission, 2009: 3).

At the summit marking the tenth anniversary of the Barcelona Process in 2005, it was decided to renew efforts to protect the Mediterranean and in November 2006[10] the Cairo meeting of Euro-Mediterranean environment ministers agreed a road map with the aim of tackling the main land-based sources of pollution by 2020. This became known as Horizon 2020. Selection of de-pollution of the Mediterranean as a priority theme for the UfM thus reinforced a commitment which had recently been given renewed emphasis within the EMP.

In early 2007 a technical steering group was established, co-chaired initially by Morocco and Finland and with membership open to all members of the UfM, and national contact points were identified. Three technical sub-groups for pollution control, capacity building and monitoring were also established in January 2007. In mid-2007 a consultancy team was appointed by the EIB to identify priority projects. Following consultation in seven potential host countries, the ensuing report selected 44 physical infrastructure projects from a list of 131 previously identified in UNEP's Strategic Action Plan for the Mediterranean. The list, with a total estimated cost of €2.1 billion (EIB, 2008: 11), was adopted at the first meeting of the pollution sub-group in March 2008 and the results disseminated to potential funders. Subsequently, a core group of official funding agencies – the European Commission, the EIB, the French Development Agency, the German Development Bank, UNEP and the World Bank – was reported to be collaborating to select projects for funding (Commission, 2009).[11] A conference on private investors' involvement in funding Horizon 2020 projects was held in Athens in October 2007. By 2008/09, seven projects from the list had entered the European Neighbourhood Investment Facility funding pipeline and, in March 2009, the EC provided grant support from the Facility for Euro-Mediterranean Investment and Partnership (FEMIP) of €800,000 to assist in project preparation when applying for International Financial Institutions (IFI) funding.[12]

Capacity building under Horizon 2020 builds on projects begun from 2003 as part of the MEDA-WATER programme. With respect to monitoring, following the 2006 Road Map agreement, the European Commission brought together the European Environment Agency (EEA) and UNEP/MAP to develop a Shared Environment Information System for the Mediterranean. A work programme was agreed by the Review, Monitoring and Research sub-group (chaired by the EEA) in October 2008, focusing on development of a biannual report and score card, and of a composite environmental indicator, plus links to research activities.

This initiative is intended to be regional, albeit with scope for sub-regional groupings for some individual projects but the aim is to develop national and sub-national (regional and local government level) capacity for physical project design and implementation. The regional elements stem from the shared benefits that derive from improved management of common property and from the need for commitment by all countries bordering the Mediterranean to participate if pollution reduction targets for the sea are to be met. For the EU, a key goal is to extend its environmental standards with respect to solid and liquid waste management and pollution prevention, through both physical investments and implementation of common monitoring measures. If effective, Horizon 2020 can be expected to bring benefits to users of the Mediterranean both as producers (e.g. the fisheries sector) and consumers (e.g. tourists).

*Maritime and Land Highways*

The key features of the Maritime and Land Highways project theme are development of motorways of the sea and coastal land motorways, modernization of the trans-Maghreb rail link and maritime security and safety. This theme too has strong EU support and potential links with other EU transport projects and it too is based on on-going programmes, both regional and bilateral: assistance for the development of transport infrastructure in partner states, especially ports and roads, have been important features of both the EMP and ENP.

In 2001, the European Commission decided, 'in the framework of the MEDA Programme', to undertake the first Euro-Mediterranean Regional Transport Project, with the aim of facilitating trade expansion both between the EU and partner states and between the latter. The project, which was designed to operate in conjunction with other regional and bilateral transport initiatives funded by MEDA, the World Bank and other international and regional organizations, ran from 2003 to 2009 and featured a number of components: a diagnostic study, development of a regional transport strategy and a Regional Transport Action Plan, plus promotion of policy dialogue and capacity building (including local capacity for project design and preparation). Projects promoted included institutional, administrative and procedural reforms relating, for example, to customs clearance, with a view to speeding up both movement of freight and turn around times for ships in port. Other MEDA transport-related regional projects included SAFEMED, designed to promote Euro-Med co-operation in the field of maritime safety, and REG-MED, designed to promote regulatory convergence to facilitate international transport in the Mediterranean. In 2006 the Motorways of the Seas Project, Phase I, was introduced as a sub-component of the Euro-Med Transport Project. Bilateral MEDA projects included development of port facilities in a number of partner economies including both Morocco and Syria. The regional and sub-regional projects listed in this paragraph continue under the UfM.

This project theme is a regional one, divisible into sub-regional components. Completion of some regional transport projects already underway prior to the UfM continues to be delayed by national political and/or economic interests (e.g. absence of completion of the cross-border links between Morocco and Algeria for the trans-Maghreb motorway), thereby reducing the overall efficiency of the regional transport network. While project implementation continues, including some port upgrading, there is no clear evidence that the protagonists of this theme – politicians, bureaucrats, firms and consultants engaged in project implementation – have succeeded in either raising its profile or raising the momentum of investment in the transport sector relative to pre-July 2008.

Integrated land and sea transport, with emphasis on short sea routes (motorways of the sea) for moving freight, can take pressure off already congested roads and help to protect the environment. However, whether all partner states give all aspects of this project theme as high a priority as the EU is open to question. In various communications and memoranda Commission staff note difficulties in mobilizing both public (ministerial) and private stakeholder interest in partner countries (see EuroMed Transport Programme, 2009: 2). This may be partly due to the range of transport-related projects already on-going or under discussion in partner states, as well as the relatively high standard of the main transport arteries in much of the region and internal difficulties with dealing with procedural reforms urgently needed to speed up movement of goods.

*Renewable Energy: The Mediterranean Solar Plan (MSP)*

Prior to 2008, there are references in various Euro-Med documents to sustainable energy but there was much stronger orientation towards promotion of market harmonization and integration for existing energy supplies: gas, oil, electricity. EIB loans chiefly facilitated completion of electricity and gas links, both North–South and South–South. Identification of renewable energy as a key component of one of the six UfM project themes thus represents a more innovative initiative than either of the themes reviewed so far. Like them, the theme entails a programme of projects, most not yet identified, which could take several decades to achieve long-term targets. From the perspective of the EU an important motivating factor is the anticipation that renewable energy imports from partner countries could help in meeting its own green energy targets.

An important feature of the MSP, which aims to develop 20GW of renewable energy production capacity and to achieve 20 per cent energy savings around the Mediterranean by 2020, is the strong private sector support for investment in concentrated solar energy production which comes from the D11 consortium (14 large corporations from the Euro-Med region plus DESERTEC (a non-profit foundation concerned with promotion of solar power generation). The Plan is analysed by Darbouche (this collection). Here, points to note concern the significance of a dynamic project theme driver and of the MSP's proposals for institutional reform in the energy sector.

In 2008–09 the MSP had a dynamic driver in the French minister for ecology, energy, sustainable development and land management, Philippe Lorec, and planning went ahead in the absence of a UfM secretariat. Following preparatory meetings on technology options, cost estimates and grid interconnections, a November 2008 conference in Paris of actors from the public and private sectors agreed the content of Phase 1 of the MSP (2009–10), to include the elaboration of procedures for developing, structuring and financing projects, the launch of pilot projects and proposals to create or reinforce a number of training centres, plus a research and development platform focused on technology development for solar energy. It was also agreed that in Phase II (2011–20) the full MSP will be elaborated and implemented (Mediterranean Solar Plan, 2009). Lorec had the backing of the Egyptian–French co-Presidency, which organized further meetings in 2010. Meanwhile, in 2009 the EU also continued to organize meetings on Euro-Med energy issues. It remains to be seen whether the Secretariat will successfully take over as the Plan's driver and/or whether the Plan will receive equally strong organizational support from the next co-Presidency.

Investments in renewable energy generation are in the first instance national, not regional, projects. A regional or sub-regional element arises when power is to be exported. This has two implications. First, national investment in expanded generating capacity for export presupposes either the existence of, or, as is needed in the UfM context a prior commitment to create, the necessary grid connections. Secondly, regulatory reform and co-ordination are needed, for several reasons: to facilitate approval for trans-national projects and to synchronize both technical standards and national incentive structures.[13] DESERTEC has identified regulatory co-ordination as the key role for the UfM in facilitating the D11 project. Institutional development lacks the visibility expected of UfM project themes but remains an essential aspect of all three physical infrastructure themes, as well as of the civil protection theme.

The scope which this project theme offers for a sub-regional approach to implementation is reflected in the June 2010 commitment by Maghreb energy ministers to electricity market integration.[14]

*Civil Protection*

Under the EMP the goal of increased security in the Mediterranean region was pursued through the economic and the political and security pillars. Civil protection itself featured in the latter. Its presence also as a priority theme of the UfM represents another instance of project evolution from previous periods. Under the Barcelona Process a five year pilot project on co-operation in civil protection was introduced in 1998, followed by a 'bridge' project which focused on information sharing and political confidence building. Four countries participated in this phase: France, Egypt, Italy and Algeria. Under the UfM, Phase 3 – Prevention, Preparedness and Response to Natural and Man-made Disasters (PPRD South) – started in March 2009 and will run to 2012. An action plan launched in October 2009 emphasizes dialogue, information sharing and capacity building in relation to national and regional risk assessment, prevention and preparedness for disasters, plus the running of a 'full-scale' regional simulation exercise of response to a major catastrophe (Faridi, 2009). The project has EU funding of €5 million and is managed by a consortium led by the Italian Civil Protection Authority. Designated beneficiaries are all UfM partners with Mauritania and Libya as observers (http://www.euromedcp.eu). A stated goal is to shift emphasis from disaster management to disaster prevention, but speakers at the plan's launch acknowledged considerable difficulty in arousing media and public interest in some aspects of disaster prevention: e.g., difficulty in generating sufficiently precise and reliable risk estimates for residents of earthquake zones impedes preventive evacuation. (The same problem is faced by the World Bank's Global Facility for Disaster Reduction and Recovery, from which this theme may receive support.)

*Higher Education and Research: The Euro-Med University (EMUNI)*

The commitment to establish the Euro-Med University coincided with the period following Sarkozy's election to the French Presidency, and the pushing of his proposal for a Mediterranean Union, but the timing appears to be coincidental. From the early 2000s dialogue among EU and Mediterranean partner country universities developed through the Euro-Med Universities Forum. A proposal to establish a dedicated Higher Education space in the region was endorsed at a Forum meeting in 2005. In 2006, the Slovenian government, in apparent anticipation of its 2008 presidency of the EU, instigated a review of options for establishing an international higher education institution. In 2007, Euro-Mediterranean culture and education ministers met and endorsed 'the establishment of a Euro-Mediterranean space of higher education and research' (Euro-Med Parliamentary Assembly, 2008: 5). Also in 2007, the Slovenian government, apparently aware of support from within the Euro-Mediterranean region, established both a secretariat with responsibility for the founding and running of the proposed university and the EMUNI Foundation with responsibility for raising finance. EMUNI offered its first programmes in 2008–09, drawing on contributions from some of its 115 member organizations located in 32 countries (mainly universities plus other higher education organizations).

The UfM has, however, impacted on EMUNI through determination of its priority teaching and research themes (plus limited funding support). EMUNI's statement of aims asserts that both teaching (at masters and PhD level) and research will be oriented towards the priority themes of the UfM plus a number of others, such as migration studies, which address issues of concern to the Euro-Med region. The foci of some of the 14 masters study programmes announced for 2009–10 are, however, indicative of the difficulty which EMUNI may face in consistently mounting programmes oriented towards UfM thematic areas as long as it lacks the core faculty to undertake course design and delivery: while the energy sector is not represented, some courses, such as 'Landscape, Territory and Patrimony' and 'Maritime Civilizations in Light of Environmental Changes', appear to be tangential rather than central to the priority themes (EMUNI, 2009: 29–30). Current funding is apparently modest: the 2009 Annual Report records funding for the 2009 programme of €0.8 million from the Slovenian government to cover establishment and start-up, €0.38 million raised by the EMUNI Foundation, European Commission funding of €1 million spread over three years for development of study programmes and planned projects, tuition fees and 'other sources from the applied projects'. Of the six UfM priority themes, EMUNI is the only one which does not have support from complementary World Bank funding activities in the region, although the Bank does support tertiary education governance and quality assurance among partner states as well as some intra-regional collaborative academic centres, including the Marseilles Centre for the Mediterranean and the Arab Water Centre.

Recent evidence suggests a possible broadening in the focus of this project theme. The programme for the May 2010 UfM funding meeting in Marseille, organized by the co-Presidency, included a Human and Social Development Workshop covering, inter alia, higher education and research: the programme lists various regional scientific and research centres, including the Mediterranean Centre for Scientific Research and the Euro-Mediterranean Institute for Risk Science (a 23 university consortium), but not EMUNI (Darwish, 2010). However, H.E. Ma'sadah, secretary-general of the UFM has acknowledged publicly the strategic role that EMUNI plays in the UFM and the university's directorate continues to develop links with higher education centres in the region.[15]

*The Mediterranean Business Development Initiative*

As conceived for inclusion in the priority themes, the Mediterranean Business Development Initiative is intended to facilitate access to credit for Mediterranean small, medium and micro-enterprises and to complement existing projects concerned with SME development in partner states, including the Invest-in-Med project (2008–11). Invest-in-Med (75 per cent EU funded) aims at developing sustainable trade relations, investments and enterprise partnerships between the two rims of the Mediterranean. Activities promoted include networking (e.g. to enable commercial and technological partnerships), assistance (including staff exchanges) and documentation (including economic intelligence reports). However, this programme does not provide credit to SMEs and the proposed Business Development Initiative is thus both innovative and complementary to it. The Initiative was promoted by Italy and Spain but at the time of the 2008 UfM summit it had apparently had little, if any, underlying planning. Subsequently, it became the focus of a feasibility study by the EIB. It then emerged that the initiative overlaps with a regional SME finance facility under development by the World Bank. Co-ordinated development of a single project is under discussion but project design is still to be finalized (C. Kenny, World Bank, personal communication, 26 July 2010). Meanwhile, as with EMUNI, the agenda for the May 2010 UfM funding meeting was suggestive of a shift in emphasis towards a broader scope for the project theme, building also on activities planned prior to the UfM, such as Invest-in-Med. This was also advocated in the concluding declaration of the Mediterranean Economic Leaders Summit of June 2010 (Barcelona Euro-Mediterranean Business Declaration, 2010: 3).

*Continuity, Change and Regionality in Project Themes*

This overview suggests (in partial contrast to Bicchi, this collection), that continuity matches or outweighs change in the choice of UfM project themes. Evidence of continuity and/or of evolution from programmes which had been promoted prior to 2008 is present in at least four of the priority themes of the UfM. While the themes all have a regional 'embrace', a number have significant sub-regional, national and intra-national elements. This, a function of both realism and political pragmatism, also reflects continuity with the Barcelona Process.

## The Project Themes: Development Potential and Implementation Viability

This section starts with an overview of the potential contribution of the priority themes to economic development among UfM members, particularly in partner states. It then considers certain aspects of the UfM's design which may impact on project theme viability and performance.

*Potential Impact of Project Themes on Economic Development*

Economic development may be interpreted in terms of outcomes – specific goals met wholly, partially or not at all – or in terms of performance in implementing a process which can in turn be expected to generate these outcomes (see above). Core goals of economic development are: sustained growth of output per person, in a manner which conserves non-renewable resources and does not put at risk the welfare of future generations, combined with an increase in employment and a reduction in poverty. A widely accepted interpretation of the process of economic development focuses on the expansion of sectors which offer scope for sustained advances in labour productivity, both through scale economies and technological advance, including, most notably, manufacturing.

The UfM physical infrastructure project themes each have potential to contribute to sustainable development: sea transport has a lower carbon footprint than road transport, the MSP is intended to provide 'green energy' both for local consumption and export to Europe and de-pollution of the Mediterranean will help to sustain both marine life and tourism. The MSP offers scope for reducing energy poverty, de-pollution of the Mediterranean may enable some expansion of the tourism sector with positive impact on employment, and transport development will facilitate trade and, through this, encourage increased competition in production. However, it is impossible to make accurate predictions of likely development impacts – in terms of both goals and process – when most component projects are yet to be implemented (and, in some cases, identified), when the time horizon for their implementation is uncertain and when different possible scenarios for output growth and trade, themselves subject to multiple influences, may impact on project outcomes. This applies both to the scale of net project benefits and how these will be distributed – between the EU and partner states, between the partner states themselves and between individuals, including the poor and unemployed. The contribution of the infrastructure projects to increased employment in the region will depend in the short term largely on the degree of labour intensity, and the scale, of capital construction and, in the longer term, on the scale and pattern of output growth and trade which they subsequently induce. With the possible exception of energy, impacts on poverty will be incidental, depending inter alia on patterns of expenditure of locally generated incomes. For example, the categories of individuals who benefit from Horizon 2020 priority projects will be determined primarily by geographic location and by user activity. They will include tourists from both sides of the Mediterranean but, over the next decade at least, chiefly from the North, and elements of the regional fisheries sector, for which the largest boats also usually do not belong to partner state nationals. Whether, and to what extent, these projects bring indirect benefits to the poor will be mainly an incidental function of the pattern of beneficiary expenditures.

With respect to the other project themes, civil protection is unlikely to make a notable contribution to growth, employment or poverty reduction, or to transformation of production processes, with the possible exception of construction technology: the theme's role is to protect given values of output, employment and incomes, and of public and private assets, in the face of disaster risk and disaster outcomes. EMUNI could contribute to productivity enhancement by helping to plug skill gaps, the degree of impact depending on the scale, focus and effectiveness of its teaching and research programmes. However, there can be a significant time lag from initial investment in training to generation of economic benefits arising from the application of new skills or research knowledge. The Mediterranean Business Development Initiative (MBDI), either with its initial stated focus on credit provision or with a broader focus embracing other SME-focused initiatives, notably Invest-in-Med, may initially have rather low and localized impact, unless there is success in funding dynamic enterprise clusters. Lack of credit, the theme's core, is one of several variables which may constrain the development of small, medium and micro-enterprises in the Mediterranean region as elsewhere. Where the credit constraint is dominant, releasing this through a new UfM/World Bank micro-finance project could indeed contribute to output growth. Whether, and over what time horizon, it would also lead to employment growth depends on whether the credit is used to fund labour saving technical change. Over recent decades there is abundant evidence of firm reorganization and capital investment resulting in 'jobless growth' in the short term, although in the longer term any consequent enhancement in competitiveness, including in export markets, can lead to further output expansion which is job creating. Indirect gains to the poor may ensue from some of the local expenditure of those workers who do benefit from output expansion.

## Implementation Viability

If UfM priority project themes are to have a significant impact, then new funding must be raised to resource an increased rate of project implementation. Since the EU is unwilling to commit additional funds to the UfM project themes over and above funds committed under the ENPI, use of private finance and public–private funding partnerships is to be given increased emphasis. There is significant private sector interest in large-scale investment in solar power – the focus of the MSP – but apart from this sector, private sector interest has so far been limited. Whether additional interest can be mobilized in the other themes, and whether the interest of official funding agencies can also be enhanced, depends partly on whether the UfM can sustain a lead role in identifying and communicating funding opportunities relating to these themes, while also facilitating funding applications by partner states (roles designated to the Secretariat but during 2008–10 played by the first co-presidency supported by the EU). Increased investment in priority themes will also depend on whether the UfM can successfully promote a more attractive investment environment through institutional reform and harmonization of regulations.

Recognition of a continuing role also for multilateral and bilateral funding agencies is reflected in the EU's commitment of resources to assist partner states in preparing funding applications. However, with a few exceptions, notably France's Agence Française de Developpement (FDA), there is little evidence that the UfM has as yet resulted in additional funds being mobilized. One of, if not the, largest potential co-funder among official development agencies is the World Bank, which has programmes either in play or under development relating to each of the priority themes. However, the Bank's long lead time from project conception, through design and evaluation to implementation, mean that to date the UfM themes have not fed through into influence on new funding commitments by the Bank. The Bank has a 'Sustainable Med' programme which pre-dates the UfM. It is preparing a Mashreq trade facilitation/infrastructure facility, to be followed with one for the Maghreb, which will enhance funding and technical support for the Maritime and Land Highways theme. It has committed $755 million to UFM partner economies under its Middle East and North Africa (MENA) Concentrated Solar Power Programme and, as noted above, intends to develop a regional SME financing facility, as well as its Global Facility for Disaster Reduction and Recovery and support for tertiary education governance and quality assurance. These common interests offer scope for co-ordination between the UfM secretariat and the Bank with respect to future commitments. So far, co-ordination has been more *ad hoc* than strategic, with Bank representatives attending UfM project planning and funding meetings which relate to their particular fields, although in the case of the MBDI there is also regular contact between the individuals at the EIB and the World Bank who have responsibility for the development of a SME finance programme for the southern and eastern Mediterranean.

The launch of the UfM gave new impetus to efforts to organize briefing meetings for potential private and public sector funders. Meetings organized by the co-Presidency include ones held in Alexandria in April 2009 and in Marseilles in May 2010. To coincide with the Marseilles meeting the formal establishment of the InfraMed Fund (first proposed a year earlier in Alexandria) was announced. The Fund will contribute to the Union for the Mediterranean by providing a source of equity financing for sustainable urban, energy and transport infrastructure projects in partner states. Initial commitments are €385 million,[16] with a target of €1 billion. Due in part to the very high capital costs of constructing rail roads and major road arteries, it is likely to be the land transport component of the Maritime and Land Highways theme which is most severely held back by lack of finance: within partner states this is a sector in which other national needs for transport infrastructure may compete for funding with the proposed major regional or sub-regional transport arteries. Furthermore, 'financial groups seem to be less interested in transport infrastructures faced with the increasing interest in energy infrastructures' (Barcelona EuroMed Forum, 2010).

*Implications of the UfM's Political Goals and Organizational and Institutional Structures for Implementation of Priority Project Themes*

One problem with raising the political profile of the UfM, discussed elsewhere in this collection, is the existence of disputes and high levels of political tension between some members; these may lead to boycott of meetings and to unwillingness to accept key roles such as that of co-president. As Bicchi shows in the introduction, the decision to raise the UfM's political status may make it harder to achieve joint commitment to technical/infrastructure projects with a regional or sub-regional focus by states between which there is political conflict. It is therefore important that levels of national self-interest in the completion of the regional and sub-regional programmes of the UfM should, where possible, be high for all intended participants – in particular with respect to interconnections of physical infrastructure and the maintenance of common property.

The Secretariat, given its composition, should promote a sense of co-ownership through the implementation of its designated roles in identifying and supporting the preparation of priority projects and in mobilizing funding and monitoring them. Whether it will be allocated the resources which are necessary for it to fulfil these functions remains to be seen. However, evidence presented in the previous section suggests that some momentum can be sustained even in the absence of the Secretariat. For Horizon 2020, the technical steering group and the EIB have fulfilled roles – project identification and seeking funding – allocated to the secretariat. The maritime and land highways theme also inherited a fairly well established planning infrastructure from the previous 'regime'. Meanwhile the first co-presidency has played a significant role in organizing meetings to brief potential funders for the project themes, and one co-president, France played an important role in co-ordinating and spearheading the development of the MSP. The EIB also has responsibility for designing the MBDI. The fact that the EU's infrastructure programmes for energy, transport and the environment include an important Mediterranean focus also suggests that, should UfM structures fail to deliver, then it will attempt to continue to drive forward project identification and implementation within these themes, including through the ENPI which, some have suggested, might be extended to include a multilateral arm (Aliboni, 2010).

**Conclusion**

The economic chapter of the EMP focused on promoting economic policy and institutional reform in partner states, supported by financial assistance intended for budgetary and balance of payments support and for funding investments in physical infrastructure perceived as necessary if partner states were to be able to respond to new competitive stimuli and to attract FDI. While the UfM's focus on physical and social infrastructure project themes stands in apparently marked contrast to the EMP's economic focus on policy and institutional reform it in reality represents reinforcement of key components of the Barcelona Process (continued under ENPI), albeit with some innovation, notably the new emphasis on green energy and the proposed MBDI. Regulatory harmonization, one of the goals of the EMP, also remains an integral component of the UfM's infrastructure themes. Where, however, the two differ is in the absence from the UfM of any emphasis on strategic policy reform.

The UfM Secretariat, which is intended to drive priority project themes, was still not fully functional by the end of 2010. Whether this matters for effective theme implementation is not yet clear. A strong driver for an agreed project within the co-Presidency, especially if backed by the EU Commission, *may* substitute for the lack of an effective Secretariat. Likewise, where organizational structures for the six priority themes are already in place due to continuity with activities initiated prior to the UfM, it is unclear whether the Secretariat can significantly add to their momentum, at least in the short term: this will require a depth of familiarity with the projects, with technical issues and with relevant contacts across a range of fields, which will not initially be fully available to the Secretariat both due to the time required for in-depth familiarization and to low staffing levels within its individual sections. With respect to new priority theme identification, if any, this could come through the co-Presidency and/or through the Parliamentary Assembly of the UfM or meetings of local government representatives, for discussion and endorsement at the biannual summits if these are maintained. The absence of a Secretariat could, however, weaken potential for co-ownership and reinforce the influence of the EU.

In reviewing the experience and outcomes of the EMP and the ENP, one striking element is the EU's relative strength, compared with partner states, in promoting change in its own interest. The EU has consistently used the EMP and, more recently, the ENP in conjunction with the UfM, to promote its own sustained development. This has not been matched by an equally coherent, strategic approach to negotiation with the EU on the part of partner states. As noted elsewhere in this collection, the institutional design of the UfM provides scope for this to change, by providing enhanced scope for the political concerns of especially the eastern Mediterranean states to over-ride progress on other fronts, thereby threatening the future of the UfM. But does this pose an equal threat to project themes? EU pragmatism, combined with its recognition of scope for re-emphasizing the technocratic element in Euro-Med relations, may ensure the continuity, either on a regional or sub-regional scale, of those project themes to which it attaches most importance, albeit with greatest success where there is genuine joint interest.

While the project themes may have a life independent of the UfM, the debate surrounding the proposals for, and the subsequent introduction of, the UfM, served to inject new life into some of them, with enhanced momentum in the organization of working groups and planning meetings both before and following the 2008 Summit as well as the establishment of co-chairmanship of some of these prior to July 2008. There is also evidence of renewed momentum in priority theme planning and efforts to promote funding in spring 2010 prior to the projected, but postponed, UfM biannual summit. So far, UfM impacts on project funding and implementation, and hence potentially on regional development, have been negligible but this is largely inevitable given the length of the time required for major new project design, for putting in place key elements of regulatory harmonization and for funding mobilization. Over the first two years of its existence, the UfM's contribution to facilitating future economic development has consequently been limited. There is potential for stronger impacts in the future in the Mediterranean region, but these are conditional on success in mobilizing funding and on the impact of intra-regional political tensions in impeding completion of some regional projects.

## Notes

[1] High oil revenues may also impede export diversification and expansion through their impact on foreign exchange rates.

[2] Throughout this paper, the phrase 'partner states' refers to the partner states of the southern and eastern Mediterranean and, with respect to the UfM, also to those of the eastern littoral of the Adriatic.

[3] Partner states had had privileged access to EU markets for manufactured exports since the 1970s.

[4] On the neo-liberal tenets in EU foreign policy, see Holden (this collection)

[5] Among the difficulties in assessing the macroeconomic impacts of the Barcelona Process are: the lack of a counterfactual, the time period for which data are available relative to that over which growth impacts may take effect and the complementary role of non-EU agencies in assisting the liberalization process.

[6] For a fuller account of Tunisia's economic performance both before and since 1995, see the Economics Intelligence Unit (1996).

[7] Many traditional family firms were initially unwilling to open their books for scrutiny.

[8] Investment facilitated partly by EU-supported upgrading of port facilities includes the 2008 Renault-Nissan alliance to establish a low cost car plant near the new port of Tangiers-Med (Economist Intelligence Unit, 2008a: 22).

[9] Examples of 2003–07 project groupings include EMWATER: Turkey, Lebanon, Jordan, Palestine, Germany, Italy; MEDAWARE: Greece, Spain, Morocco, Turkey, Cyprus, Lebanon, Palestine, Jordan; ZERO-M: Egypt, Morocco, Turkey, Tunisia, Austria, Germany, Italy. See http://www.emwis.net/initiatives/fo1060732/proj (accessed 27 November 2009).

[10] This was two months before Sarkozy was selected as a candidate for the French presidency.

[11] In 2006, France's Agence Française de Développement (AFD) committed €230 million to finance de-pollution projects in Morocco, Tunisia and Egypt, followed in 2008 by commitment of a further €500 million for de-pollution projects in partner countries.

[12] Following entry of the coastal Balkan states into the UfM, it is intended to conduct a similar exercise for them and Turkey, using these countries' National Action Plans (prepared as part of an earlier UNEP initiative).

[13] In the EU, prices for imported solar power need to be raised if investment in export from UfM partners is to become viable (C. Kenny, World Bank, personal communication, 26 July 2010).
[14] Source: http://desertec-mederranee.over-blog.com/article-experts-meeting-on-the-mediterran (accessed 3 November 2010).
[15] A memorandum of understanding was signed in Beirut with the Association of Arab Universities also in May 2010. See http://www.emwis.org/thematicdirs/eflash/flash80 (accessed 30 October 2010).
[16] Caisse des Dépôts (France) and Cassa Depositi e Prestiti (Italy) will each commit €150 million, Caisse de Dépôt et de Gestion (Morocco) €20 million, EFG Hermes (Egypt) €15 million and the EIB €50 million (EIB, 2010).

## References

Aliboni, R. (2010) *The Union for the Mediterranean: Evolution and Prospects*, Istituto Affari Internazionali, Rome, Document 0939E, updated February 2010.

Barcelona Euro-Med Forum (2010) *Union for the Mediterranean: Projects for the Future*, May. Available at http://www.slideshare.net/soniabess/union-for-the-mediterranean-projects-for-the-future (accessed 3 November 2010).

Barcelona Euro-Mediterranean Business Declaration (2010) *Taking the Initiative, Shaping the Union for the Mediterranean*, June. Available at http://www.ueapme.com/IMG/pdf/100604–JointDeclaration EuroMedBarcelona_FINAL_EN.pdf (accessed 28 July 2010).

Commission of the European Communities (2009) *A Progress Report on the First Three Years of Horizon 2020*, Commission Staff Working Paper, SEC(2009), 1118 final, Brussels, 6 August.

Darwish, R. (2010) *For'UM Union pour la Méditerranée*, Global Arab Network, 14 May. Available at http://www.wnglish.globalarabnetwork.com/201005145876/Economics/egypt-france (accessed 21 July 2010).

Economist Intelligence Unit (EIU) (1996) *Country Profile, Tunisia, 1996–97*.

Economist Intelligence Unit (EIU) (2006) *Country Profile, Tunisia, 2006*.

Economist Intelligence Unit (EIU) (2008a) *Country Profile, Morocco, 2008*.

Economist Intelligence Unit (EIU) (2008b) *Country Profile, Tunisia, 2008*.

Economist Intelligence Unit (EIU) (2008c) *Country Profile, Jordan, 2008*.

Euro-Mediterranean University (EMUNI) (2009) Annual Report. Available at http://www.emuni.si/Files//Denis/Brochures/Annual _Reports/EMUNI_Report_2009.pdf (accessed 7 December 2009).

Euro-Med Parliamentary Assembly (2008) *Recommendation for the March 2008 Athens meeting of the Assembly*, Working Group on the establishment of a Euro-Mediterranean University. Available at http://www.eppgroup.eu/euromed/.../080327ATHENES_cultural_committee_EN.pd accessed (6 December 2009).

Euro-Med Transport Project (2009) *MEDA Bilateral Projects* and *MEDA Regional Projects*. Available at http://www.euromedtransport.org/6.0html (accessed 14 December 2009).

Europa (2009) Summaries of EU legislation, *Mediterranean Partner Countries, MEDA programme*. Available at http://europa.eu/legislation_summaries/external_relations/relations_with_third_countr... (accessed 3 November 2009).

Europa (2010) Summaries of EU legislation, *The Barcelona Declaration*. Available at http://europa.eu/legislation_summaries/external_relations/relations_with_third_countries/mediterranean_partner_countries/r15001_en.htm (accessed 28 April 2010).

European Investment Bank (EIB) (2008) *Horizon 2020 – Elaboration of a Mediterranean Hot Spot Investment Programme (MeHSIP)*, Final Report for Contract REG/2006/02, January.

European Investment Bank (2010) *As a Major Contribution to the Union of the Mediterranean: InfraMed Infrastructure Fund is Launched*, May. Available at http://www.eib.org/about/press/2010/2010-078-lancement-du-finds-dinfrastructure (accessed 21 July 2010).

European Parliament (2007) *Resolution of 21 June 2007 on MEDA and Financial Support to Palestine*. Available at http://domino.un.org/unispal.nsf/59c118fo65c4465b852572a500625fea/7abdc950e26e... (accessed 3 November 2009).

Faridi, N. (2009) *Civil Protection Cooperation in the Mediterranean Takes Action*, 23 October. Available at http://eurojar.org/en/euromed-articles/civil-protection-cooperation-mediterranean (accessed 5 December 2009).

Khan, S. (2010) Exploring and naming an economic development alternative, in: S. Kahn & J. Christiansen (Eds) *Towards New Developmentalism* (Abingdon: Routledge).

Mediterranean Solar Plan (2009) Global Presentation. Available at http://www.slideshare.net/soniabess/MSP/-global-presentation (accessed 15 December 2009).

Morin, O. (2005) Le partenariat euro-mediterraneen, *Etudes*, February. Trans. by D. Macey, in: *Revue des revues de l'adpf, selection de juillet 2005*. Available at www.diplomatic.gouv.fr/fr/fr/IMG/pdf/0103-Morin-Z.pdf

SEMIDE EMWIS (2006) *MEDA Figures*, 16 January. Available at http://www.emwis.net/overview/fol101997/fol221357/doc988673 (accessed 11 November 2009).

World Bank (2009) *World Development Indicators Database, September, 2009*. Available at http://web.worldbank.org/WBSITE/EXTERNAL/DATASTATISTICS (accessed 16 April 2010).

# Third Time Lucky? Euro-Mediterranean Energy Co-operation under the Union for the Mediterranean

HAKIM DARBOUCHE
Oxford Institute for Energy Studies

ABSTRACT *Energy co-operation has appeared as a priority area on the EU's Mediterranean policy agenda since the promulgation of the Euro-Mediterranean Partnership (EMP) in 1995. The Union for the Mediterranean (UfM) has pledged to do more than previous policy frameworks in the area of energy co-operation, and has specifically identified renewable, particularly solar, energy as a possible catalyst. This study aims to assess the prospects for Euro-Mediterranean energy co-operation within the framework of the UfM. To this end, it will examine the reasons behind the failure of both the EMP and the ENP to achieve meaningful progress in their equally sanguine enunciated policy objectives in this area, comparing their respective approaches with that of the UfM. It will be argued that the prospects for Euro-Mediterranean energy co-operation under the UfM will hinge more on the shifting priorities of European consumers and SMC producers, informed in particular by concerns over climate change, the need to diversify sources of primary energy supply and the depletion of proven conventional fossil fuel reserves, than on the attributes of the UfM per se.*

European concerns over energy security have invariably been fuelled by Europe's dependence on external sources of oil and gas supply to meet its energy needs. Today, the European Union is the world's biggest importer of primary energy and its second largest consumer after the United States. As this dependence is projected to grow steadily in the coming years[1] – even in light of the adjusted demand forecasts imposed by the economic recession – a sense of urgency amongst EU policy makers has put energy security at the heart of recent EU foreign policy initiatives. Indeed, a number of global and interrelated energy market trends have highlighted the need for a strong and coherent EU external action to complement its internal energy security policy agenda. These include the concentration of conventional fossil fuel reserves in and around a handful of unstable countries and regions of the world; growing global demand and competition for access to these reserves, particularly amongst the emerging economies of Asia and the Middle East; rising global energy prices; and the perceived politicization of energy trade by some major producers.

Europe's sense of energy vulnerability was exacerbated in recent years by the assertive course of action adopted by its single most important supplier of oil and gas, Russia. The revival of the energy industry in Russia since the late 1990s coincided with reconfigurations of domestic and international power relations, leading to a review of foreign policy priorities, including in the energy realm, on the part of Moscow. This renewed determination set Russian policy makers on a collision course over energy relations with their counterparts in Ukraine and Belarus notably, and, through both these important transit countries, in EU countries. When Russia's Gazprom decided in January 2006 and again in 2007, 2009 and 2010 to cut gas supplies to these countries over pricing and transit disputes, causing disruptions to the flow of gas to EU consumers, European political and industry leaders woke up to the most sobering alarm signals regarding the EU's energy import dependence.

In their corresponding strategic rethink of energy policy, EU policy makers pointed unsurprisingly to the increasing dependence on imports from unstable regions as a serious risk, and identified the need for diversification of suppliers as a key component of the more 'coherent external energy policy' that the Community needed to deal with new energy challenges. In this context, the strategic importance of North African fossil fuel supplies to Europe's growing demand and anxiety became increasingly evident. North African energy exporters (Algeria, Libya and Egypt) had hitherto played a crucial role in the development of European energy markets, particularly gas, and enjoyed what is characteristically described as stable and reliable energy relations with EU member states. More recently however, louder recognition of their potential regional role in energy security has been voiced in EU strategic policy deliberations, with officials in Brussels asserting rather optimistically that by 2013 the southern Mediterranean could be as important for Europe as Russia in terms of energy supply. This renewed interest is reflective not only of the EU's sense of urgency, but also of the fact that North Africa's growing hydrocarbon resource-base remains relatively underexplored, particularly in Libya, which has recently returned to international normality.

Yet, despite the complementarity and interdependence underpinning Euro-Mediterranean energy relations, efforts to institutionalize regional market structures have remained subdued. While earlier European initiatives such as the Euro-Arab Dialogue and the Global Mediterranean Policy, introduced in the 1970s (partly) in response to the first oil crisis, failed to initiate meaningful regional energy co-operation mainly as a result of the EC's institutional shortcomings, more recent attempts since 1995 have generated little more political engagement in regional energy issues despite their improved configurations and loftier ambitions. Both the Euro-Mediterranean Energy Partnership (EMEP) – part of the broader Euro-Mediterranean Partnership – and the European Neighbourhood Policy (ENP) have had a conspicuously hard time in their attempts to institutionalize energy dialogue between European consumers and southern Mediterranean producers, and to lead to the approximation of normative and strategic energy priorities on both sides.

Indeed, regulatory frameworks north and south of the Mediterranean remain far from convergent on important issues such as liberalization, and major energy decisions are often perceived to lead to sub-optimal outcomes in terms of regional energy co-operation and integration in the Mediterranean.

However, the UfM seems to have introduced a novel approach to Euro-Mediterranean relations, which extends to the energy sector. Architects of the new EU Mediterranean policy initiative have, alongside new institutional arrangements (see Bicchi, this collection; Holden, this collection; Johansson-Nogués, this collection), placed technical co-operation and 'concrete projects' at the heart of their enterprise, hoping to circumvent through a more functional approach (see Bicchi, this collection) the endemic macro-political obstacles that have traditionally impeded the advancement of co-operation in the region. More specifically, the Mediterranean Solar Plan (MSP) has been proposed as the key component of the UfM's energy dimension, with the aim of capitalizing on the region's renewable energy potential to foster sectoral co-operation.[2] How straightforward will the implementation of this project be, and how likely is it to succeed in serving as an impetus for regional energy co-operation?

This study aims to assess the prospects for Euro-Mediterranean energy co-operation in the context of the UfM, focusing on the medium- to long-term outlook. To do so, it will revisit the main components of energy co-operation under both the EMP and the ENP, highlighting the main differences between all three initiatives as well as their strengths and weaknesses. In doing so, it will focus on the EU's relations with the main petroleum-exporting Southern Mediterranean Countries (SMCs) of Algeria, Libya and Egypt, as it has traditionally been with producing countries that a meaningful level of regional energy co-operation proved difficult to attain (Escribano, 2010). This contribution will argue that the UfM's energy co-operation component is likely to yield more concrete results than its predecessors, but that it will owe its relative success to the stakeholders' shifting priorities as imposed by global energy trends (climate change, growing domestic energy requirements in southern Mediterranean producers and 'peak oil' concerns), more so than the attributes of the UfM per se. Finally, in line with the framework of this collection, some conclusions on the functionalist emphasis of the UfM will be drawn.

## Energy Co-operation under the EMP

Differences between the EC member states as well as inadequate Community policies continued to militate against effective European co-operation on Middle Eastern energy security issues until the early 1990s,[3] despite the concerns triggered by the policy decisions of the region's producers in the 1970s and the US's corresponding push for the creation of a strong energy consumers' front as a counterweight to the Organization of Petroleum Exporting Countries (OPEC). It was not until the EMP project entered its conception phase (1993–95) that institutionalized energy co-operation was concretely envisaged by the EU and formally proposed to SMCs. The new European approach to Euro-Mediterranean energy relations was encouraged not only by a broader, more favourable regional political context, but was also a by-product of the incipient debate on post-Maastricht EU energy policy (see European Commission (1995a, 1995b).

In the run-up to the 1995 Barcelona conference, relevant EU pronouncements explicitly identified energy as an area of interdependence, interest and necessary policy action in the Euro-Mediterranean context (European Commission, 1994, 1995c). In similar vein, the Barcelona Declaration (2005) 'acknowledge[d] the pivotal role of the energy sector in the economic Euro-Mediterranean partnership' and set out to 'strengthen [regional] co-operation and intensify dialogue in the field of energy policies'. However, though evidently important in the eyes of the architects of the Barcelona Process, energy co-operation was not considered as primary and immediate an objective of the fledging EMP as were issues relating to the Middle East peace process, migration, security and political reform.[4] This relative lack of emphasis was certainly caused by the slack petroleum market conditions prevailing at the time. Nonetheless, nominal recognition of the importance of energy as an area of regional co-operation reflected the same long-term outlook within which the new overall EU Mediterranean strategy was nested, pertaining in this particular area to the importance of the region not only in terms of hydrocarbon reserves (Tables 1, 2), but also as a transit point for energy supplies from places like the Gulf and the Caucasus. The latter aspect meant that political and socio-economic stability in SMCs was all the more essential from an EU energy security perspective.[5]

Following up on the tone set by the Barcelona Declaration and other Euro-Mediterranean energy meetings before it,[6] the European Commission put forward a programme of action for the implementation of the priorities of the proposed Euro-Mediterranean energy partnership (European Commission, 1996). The centrepiece of this programme was to be a regional energy forum of European and SMC energy officials,[7] which has since become the main vehicle of energy co-operation in the Mediterranean, along with the Euro-Med energy ministerial conference. The main function set by the EU for the Euro-Mediterranean Energy Forum (EMEF) has been to facilitate the reform of the energy regulatory and legislative frameworks of Mediterranean partners as well as their relevant industries, with a view to developing consistent policies and pave the way for more investment and the eventual integration of energy markets in the region. These were to constitute the main objectives of the periodic action plans which the EMEF regularly proposed since its official establishment in 1997.

Over the course of the following decade, three action plans were adopted by Euro-Med energy ministers within the context of the EMEP. The first blueprint covered the period 1998–2002; the second concerned the period 2003–06; and the more recent one deals with 2008–13. The underlying priority objectives of these work programmes build upon the normative and strategic foundation outlined in the second meeting of the EMEF in Grenada in 2000. Though the focus of specific action plan priorities varied slightly, their guiding objectives have remained the convergence of the energy policies of the EU and the Mediterranean partners, the integration of the Mediterranean energy markets and strengthening competition

**Table 1.** Mediterranean natural gas data (in billion cubic metres)

| | 1995 | | | 2000 | | | 2005 | | | 2009 | | |
|---|---|---|---|---|---|---|---|---|---|---|---|---|
| | Reserves | Production | Exports | Reserves | Production | Exports | Reserves | Production | Exports | Reserves | Production | Exports |
| Algeria | 3,700 | 59 | 38 | 4,520 | 84 | 64 | 4,500 | 88 | 65 | 4,500 | 81 | 53 |
| Egypt | 650 | 12.5 | 0 | 1,430 | 21 | 0 | 1,900 | 42.5 | 11 | 2,190 | 63 | 18 |
| Libya | 1,310 | 6.3 | 1.5 | 1,310 | 6 | 0.8 | 1,320 | 11.3 | 5.4 | 1,540 | 15 | 10 |
| Rest of the Mediterranean* | 540 | 21 | 0 | 440 | 21 | 0 | 400 | 16.6 | 0 | 180 | 16 | 0 |

*Notes:* Production and exports per annum. *Reference mainly to Syria, Italy and Tunisia.
*Sources:* BP Statistical Review of World Energy (2010) and OPEC Annual Statistical Bulletin (2010).

**Table 2.** Mediterranean crude oil data (in million barrels)

|  | 1995 | | | 2000 | | | 2005 | | | 2009 | | |
|---|---|---|---|---|---|---|---|---|---|---|---|---|
|  | Reserves | Production | Exports | Reserves | Production | Exports | Reserves | Production | Exports | Reserves | Production | Exports |
| Algeria | 10,000 | 1.3 | 1.1 | 11,300 | 1.6 | 1.4 | 12,300 | 2 | 1.8 | 12,200 | 1.8 | 1.5 |
| Egypt | 3,800 | 0.9 | 0.4 | 3,600 | 0.8 | 0.1 | 3,700 | 0.7 | 0 | 4,400 | 0.7 | 0 |
| Libya | 29,500 | 1.4 | 1.1 | 36,000 | 1.5 | 1 | 41,500 | 1.8 | 1.3 | 44,300 | 1.65 | 1.1 |
| Rest of the Mediterranean* | 3,800 | 0.6 | 0 | 3,600 | 0.6 | 0 | 4,400 | 0.6 | 0 | 4,000 | 0.6 | 0 |

*Notes*: Production and exports per day. *Reference mainly to Syria, Italy and Tunisia.
*Sources*: BP Statistical Review of World Energy (2010) and OPEC Annual Statistical Bulletin (2010).

within them, and the promotion of renewable energy sources in the framework of sustainable development (see Council of the European Union, 2003). For instance, the priorities of the current EMEP action plan are 1) ensuring the improved harmonization of energy markets and legislations and pursuing the integration of energy markets in the Euro-Mediterranean region; 2) promoting sustainable development in the energy sector; and 3) developing initiatives of common interest in key areas, such as infrastructure extension, investment financing and research and development (Council of the European Union, 2007).

Concretely, the EMEP has over this period seen the allocation of some €55 million by the European Commission in support of a number of regional projects aimed at contributing to the realization of the partnership's goals.[8] The European Investment Bank (EIB) has also provided some €2 billion in loans to support energy infrastructure priority projects, notably to complete electricity and gas links in the Mediterranean (European Commission, 2006a). These include the Arab Gas Pipeline, the Medgaz pipeline, the Gas Interconnection Turkey–Greece–Italy, and various North–South and South–South electricity interconnections. Though not entirely insignificant, the financial aspect of the EU's commitment to energy co-operation in the Mediterranean remains insufficient considering on the one hand the strategic importance of some of the projects and on the other the considerable financial involvement in these very projects of industry actors, national governments and international financial institutions.

Surprisingly, the EU's approach to Euro-Mediterranean energy co-operation within the framework of the EMEP seems to have evolved relatively little. If anything, its commitment to promoting collaboration with SMCs in this strategic area appears to have at best stagnated. Strategic EU policy pronouncements on the direction of the EMP at different phases of its life-cycle either placed less emphasis on energy issues than previous avowals or simply reiterated their content, suggesting formal satisfaction with the results achieved hitherto. The European Commission's (2000b) communication on 'Reinvigorating the Barcelona Process' after five years of its promulgation paid more attention to issues relating to the Middle East peace process, the negotiation and ratification processes of association agreements, stalling reforms in SMCs, and the visibility of the Partnership at grass-root levels. Energy only received passing mention alongside other sectoral issue areas such as transport, telecommunication and water. But in an attempt to duly reposition energy in this apparent pecking order, the EC outlined the following year its plans to enhance Euro-Med energy co-operation so as to take account of the fact that the energy sector in SMCs required 'radical adjustments' (European Commission, 2001). However, this appeared to be a half-hearted effort on the part of the EU as it only amounted to a restatement of the existing priorities of the EMEP. By contrast, broader EU policy declarations such as the Common Strategy on the Mediterranean and EU Strategic Partnership with the Mediterranean and the Middle East barely acknowledged the importance of energy for Euro-Mediterranean relations and the need to develop a more adequate co-operation framework (Council of the European Union, 2000, 2004).

The attitude of energy-exporting SMCs towards the EU's proposed energy partnership was understandably welcoming at first.[9] Algeria, for instance, sought to highlight both its existing energy role and potential by way of carving out a meaningful partnership with the EU and assume leadership among SMCs in the context of the prospective EMEP – all in order to break its debilitating international isolation at the time. Other SMCs like Egypt had a genuine 'apolitical' interest in such a partnership, which it perceived as a source of technical and financial support for its emerging gas industry and growing power sector. However, the interests of the gas-exporting SMCs in particular were somewhat betrayed by the EU's Gas Directive, which was introduced in 1998 and entered into force two years later for the purpose of liberalizing the internal gas market. The ensuing grievances expressed by the EU's partners related not only to the perceived unfairness and discrimination to producers inherent in European gas liberalization,[10] but also – and most emphatically – to the uncooperative approach adopted by the European Commission in the formulation and implementation of the new legislation.

Unsurprisingly, shortly after the coming into force of the Directive, the most influential gas exporters on international markets (including Algeria, Egypt and Libya) formed the Gas Exporting Countries Forum (GECF), which has subsequently become pejoratively known as the 'gas OPEC', in order to defend their interests (Hallouche, 2006; Darbouche, 2007). In its second meeting in Algiers in 2002, the GECF was used as a platform by Algerian officials to pillory the EU for failing to consult and co-ordinate its liberalization policy plans with southern Mediterranean gas suppliers (Aïssaoui, 2002). Subsequently, the EMEP became quasi-moribund as it noticeably failed to promote dialogue between and reconcile the interests of producers and consumers, as was demonstrated by Algeria's refusal (until 2007) to agree to abolishing destination clauses[11] from its existing gas supply contracts with European customers. By then, the EMP's energy dimension seemed to be slowly but surely phased out and replaced by the relevant provisions of the ENP.

## The ENP and Energy Co-operation in the Mediterranean: More of the Same

The ENP was introduced at a time of intensified debate on energy policy within the EU. The policy issues and priorities raised by the Commission's Green Paper of 2000 on energy security (European Commission, 2000a) were thrown into sharper focus by the deteriorating geopolitics of global energy markets, which manifested themselves in the debates provoked by a steep rise in oil prices from 2003 and the increasingly tense EU–Russia energy relations from 2004. Yet in proclaiming the birth of the post-enlargement policy framework aimed at governing its relations with new and existing neighbours, the EU placed a relatively timid emphasis on the importance that energy co-operation should have in this context (see European Commission, 2003a). This exercise conveyed the impression that the EU had ducked an opportunity to respond to its own calls for the need 'to speak with one voice' on external energy matters. But, undoubtedly, the novelty of the ENP's philosophy needed justifying and explaining for an overwhelmingly sceptical neighbourhood audience, and the EU's first pronouncement on the new policy seemed to have been devoted to this purpose.

However, no sooner had the ENP been officially launched in March 2003 than the EU made public its plans for 'the development of an energy policy for an enlarged EU, its neighbours and partner countries' (European Commission, 2003b). In this new policy declaration, which was clearly inspired by the 'Wider Europe' communication, the EU set out to build a 'wider energy community' predicated on its own *acquis*. The assertion underlying this overarching goal is that energy security can be achieved by 'the EU extending its own energy market to include its neighbours within a common regulatory area with shared trade, transit and environmental rules' (see European Commission, 2006b). As a result, the tone was set for the EU's incipient energy policy: market-based provisions would form the bedrock of the European approach to energy security, straddling the internal and external dimensions of a wider, common regulatory area. This was subsequently identified as an exercise in 'normative reproduction' on the part of the EU, aiming at developing a pattern of international co-operation in a given issue area (energy) based on existing modes of internal co-operation in the same sector. Subsequently, energy security concerns were seen as the main – if not the only – factor according the ENP a coherent rationale by linking its diverse geographical regions, especially considering the idea of an 'ENP energy treaty' that Commissioner Ferrero-Waldner floated during her tenure (Youngs, 2009: 24).

As the implementation of the ENP progressed, further details emerged on the way energy co-operation with SMCs was going to be carried forward. EU documents articulating the ENP's strategy and how to strengthen it (European Commission, 2004, 2006c) spelled out the 'new' plan for doing so while reiterating the value of the existing EMEP. Besides highlighting the normative and regulatory dimension that is so prominent in the existing regional framework and that will be reinforced by the proposed Action Plans, the ENP aimed to reinforce energy networks and interconnections between the EU and its partners. This meant identifying and supporting strategic infrastructure such as liquified natural gas (LNG) terminals, gas pipelines and electricity interconnections, as well as intensifying co-operation on issues of energy efficiency and technological innovation. More specifically, the idea of a 'Mediterranean energy ring', linking together SMC markets with each other and with the EU, was given a fresh impetus under the ENP – at least on paper.

Furthermore, from 2007 a flurry of new EU energy policy deliberations, which had been stimulated by the Commission's Green Paper of 2006 (see European Commission, 2006d) and the series of controversial energy disputes between Russia and its CIS neighbours, added to the already heavy battery of energy policy measures that had been promulgated by now. These new energy packages, referred to as the EU Strategic Energy Reviews, proffered as additional priorities the establishment of enhanced energy relations with southern Mediterranean producer and transit countries, as well as the facilitation of new transit routes for natural gas supplies (see European Commission, 2007, 2008b). As a result, the European Commission has offered to conclude 'strategic energy partnerships' with these countries, stressing the importance of doing so with Algeria, Egypt and Libya. Thus far only Egypt has signed (in 2008) a memorandum of understanding on such a strategic partnership with the EU.[12]

Algeria, for its part, is still officially negotiating a similar agreement, but in reality Algerian policy makers see little value in signing a standalone energy agreement with the EU, especially if it is to be based on Brussels' narrow conception of strategic energy relations. The aim is to wrest important concessions on other issue areas, and to use energy as a springboard for broader strategic co-operation. With regard to alternative gas transit routes, the EU is now betting on far-fetched projects such as the Nabucco, Trans-Saharan and Arab gas pipelines to diversify its energy supply routes and sources. While the diversification potential of these projects, which are at different stages of their development, is significant for the EU's energy security strategy, their feasibility and potential contribution to regional integration in the South are far from being a foregone conclusion.

The proposition of strategic partnerships in energy with SMCs can be read as an implicit recognition of the limited contribution of the ENP to Euro-Mediterranean energy co-operation. Indeed, the ENP's main instrument, the Action Plans, proved insufficient as vehicles of its agenda in this regard – an agenda that represented little more than the EMEP's programme. The fact that two of the most important SMC energy suppliers of the EU – Algeria and Libya – refused to take part in this policy has constituted a serious setback for the ENP's Mediterranean energy ambitions. As a result, only a small number of the enhanced normative and infrastructural energy objectives of the ENP were realized.

Moreover, the big North African energy producers tend to be sceptical of the EU's predilection for a market-oriented approach to its energy relations with the southern Mediterranean. Algeria for instance has seen its protracted efforts to reform its energy sector receive little or no support from the EU, at a time when the government needed all the support it could obtain to help it deal with fierce domestic opposition to such plans. In the end, the liberalization programme of the upstream hydrocarbon sector was abandoned, but EU policy makers were the first to subsequently criticize the unattractiveness of investment terms in Algeria. What is more, the difficulties encountered by SMC energy companies like Algeria's Sonatrach in seeking to operate in the EU's purportedly transparent and competitive internal market represent another discrediting factor of the EU energy policies. Sonatrach's misadventures since 2007 with the Spanish political and industry authorities, which sought to curb its growing commercial ambitions in the Spanish downstream gas and electricity market, represent a blatant example of the inconsistency of the EU's policies.[13] The failure of the European Commission to stand by its market rules in this instance by adopting an unambiguous stance on the Sonatrach–Spain dispute and its contrasting obsession with Sonatrach's co-operation with Gazprom led to a reinforcement of the Algerian and other SMC governments' distrust of EU overtures in this sector (Darbouche, 2007). This translates the failure of more than a decade of Euro-Mediterranean energy co-operation based on EU-led policies such as the EMP and the ENP to bring northern and southern Mediterranean interests closer.

## The UfM's Energy Dimension: EU Novelty or *coup de chance*?

Energy co-operation appeared to drive the EU's latest foreign policy initiative in the Mediterranean from the days of its original French design as a 'Mediterranean Union'. Indeed, in the hyperbolic description of his vision for the new regional initiative, President Sarkozy often referred to a reinforced Franco-Algerian co-operation as potentially the main driving force behind the proposed Union, just like the more tangible Franco-German friendship allowed in a recent past the construction of the European Union. In this vein, Sarkozy envisioned a strategic 'alliance' between state-owned Gaz de France (GDF) on the one hand and Algeria's national oil and gas company Sonatrach on the other.[14] This, it was suggested, would lay the ground for a new form of energy partnership in the region, which would see Algerian natural gas reserves 'exchanged' for French civil nuclear expertise. However, much like its sardonic political umbrella of a Mediterranean Union, this proposal soon ran into the sand, giving way to less vacuous ideas. Subsequently, the Mediterranean Union was diluted into a Union for the Mediterranean and the Sonatrach–GDF partnership was limited to the extension and reworking of LNG supply contracts to 2019.[15]

The UfM's added value is said to reside in its institutional structure, which was designed with a view to paving the way for more meaningful co-ownership of the initiative, and, equally importantly, its focus on concrete projects that would make more tangible Euro-Mediterranean co-operation for constituencies in the region. Of the six priority projects identified by the leaders of the participant countries in the Paris summit of 13 July 2008 and reaffirmed by their foreign ministers in their meeting in Marseille a few months later,[16] the development of renewable, mainly solar, energy through a Mediterranean Solar Plan has been earmarked as a catalyst for Euro-Mediterranean energy co-operation. Soon the MSP became the flagship project of the UfM, as the momentum it has gathered since summer 2008 has outpaced all other aspects of the new framework's agenda, including the setting up of a secretariat that has been mandated to implement the identified priority projects. Between 2008 and 2009, three official and many other non-official meetings took place to define the contours and elaborate proposals for the implementation of the plan. And a clear operational roadmap for the period 2009–20 has been put in place.[17]

The basic concept underlying the MSP is the development of renewable, mainly solar, energy systems for power generation in the Mediterranean, with the aim of reaching a capacity of 20 gigawatts/year by 2020 and creating a 'Euro-Mediterranean green electricity market'.[18] The solar and renewable energy potential of SMCs has been well documented for years but the idea of harnessing it to create a new form of energy trade in the region seems to have only recently received enough political support. The discourse of EU officials in support of the MSP has clearly been articulated around the political and economic threat that structural overdependence on Russian energy supplies poses for Europe, underlining the need to diversify energy supplies and meet the commitments of the European Energy and Climate Package (see Ferrero-Waldner, 2009a). Besides this geopolitical thinking, the EU also sees an altruistic contribution through the MSP for the socio-economic development of SMCs (employment generation, technology transfer and satisfaction of growing domestic energy requirements) as well as for the welfare of our planet, making the MSP a 'win–win–win' venture (Ferrero-Waldner, 2009b).

What has allowed the rapid take-off of the MSP is a confluence of political enthusiasm for the project with industrial endorsement on both sides of the Mediterranean. In fact, the MSP, as the UfM's political and institutional energy co-operation project, has an industrial twin in the DESERTEC Initiative which was launched in Munich by a consortium of European companies on the same day as the UfM's founding Paris summit.[19] Sharing the same concept as the MSP, the Desertec project aims to raise over $400 billion and supply, from North African deserts, up to 15 per cent of Europe's electricity requirements by 2050. The deployment phase of this ambitious Mediterranean solar project is expected to start in 2011/12 and so far a number of outside investors (international financial institutions (IFIs), private and sovereign funds, commercial banks) have expressed an interest in the enterprise.

Judging from the interest that the MSP has generated in various decision-making spheres, it appears that the EU has finally got its hands onto a formula that will not only allow it to foster the Euro-Mediterranean energy partnership it has long aspired for, but one that will also contribute to its energy security. However, the assumptions behind these ambitious projects are for the most part untested in the market and, unless they firm up, the MSP will either remain a 'desert dream' or only materialize in a form that will be insignificant in the face of the impending energy-related challenges that the region faces.

To start with, an adequate legal, regulatory, institutional and organizational framework will need to be set up in the target SMCs and synchronized to the extent possible with European rules to allow the development of solar-based power generation capacity on the desired scale. This will also be necessary if SMCs are to benefit from their solar power on a commercial basis, as the liberalization of their current tariff policies is a prerequisite for the development of solar energy for domestic consumption. Furthermore, the delivery at competitive prices of solar-generated power from SMCs to Europe seems presently far-fetched considering the investment it requires not only in submarines interconnections, but also in intra-European transmission capacity. The EU may be prepared to pay a (political) premium for 'cleaner' energy supplies – though not so enthusiastically in the current times of budgetary austerity – but external sources of project finance are likely to shy away from investing if the commercial viability of the MSP remains elusive.[20] Thus, the EU will need to show more ingenuity in its approach to Euro-Med energy co-operation within the UfM if it is to capitalize on the interest its plans have so far generated on the back of a positive conjuncture.

**Assessment and Prospects**

The EU's approach to energy co-operation in the Mediterranean has been predicated on an unrelenting belief in the added value for energy security of the spread to SMCs of European market rules. Both the EMEP and the relevant energy components of the ENP were guided by this philosophy, which was reinforced in the context of the latter framework as the EU's energy policy activism became imbued with a sense of urgency in the face of record high energy prices and growing Russian foreign policy assertiveness. Despite the timid record of the EMP in fostering meaningful Euro-Mediterranean energy co-operation, the pertinence of this market-based approach in the eyes EU energy policy makers seemed to strengthen in the 2000s. This arguably occurred to the detriment of Euro-Med energy relations, as petroleum-exporting SMCs saw little value in the content of the EU's overtures and engaged with them only half-heartedly.

Conceptually, it is widely recognized that the European energy policies in the Mediterranean represent(ed) an exercise in external projection of the EU's constituent norms, which is inherent in its international actorness. The EU aims to export to neighbouring countries as many of its norms as politically and economically feasible not only because this reflects its 'inner self', but also because of the functional design of such approach. Market liberalization is considered the EU's 'most potent negotiating tool in international energy interactions'. By purportedly opening up its own market, the EU hopes to gain greater investment access in third producer countries and undercut the perceived perilous implications of bilateral deals between member states and Europe's handful of main energy suppliers (Youngs, 2007: 5). The consolidation of internal market liberalization is also a way for Brussels to seek to influence the foreign policies of member states by making a number of desired substantive reforms, such as the banning of destination clauses, inescapable. This ensures the gradual convergence of member states' foreign policies around a unified set of guiding principles, at least in relation to pivotal issues. Finally, a rule-based approach to energy co-operation is for the EU a way of contributing to the erection of well-functioning world energy markets and enticing other global powers away from geostrategic deal-making propensities (Youngs, 2009: 30).

However, little enthusiasm has been shown by a number of SMCs for the EU's offered model of extending its own market regulatory norms as a basis for energy co-operation. The EMEP and the ENP appear to have had little traction, especially in the major SMC energy producers. The reason for this lack of interest on the part of southern Mediterranean partners is two-fold. Firstly, they perceive the EU approach as 'prosaically narrow and obsessed with rules and regulations, whereas their expectation is of a co-operation model that is explicitly more strategic'. Little value do the EU's proposed co-operation frameworks add to the existing bilateral deals that member states are prepared to conclude on the basis of more strategic dividends for SMCs. Why should Algeria, for instance, replace a strategic energy partnership with France or Spain with a deal with the EU when the former allows it to obtain concessions from these member states on bilateral and other political issues as well as on broader economic matters?

Secondly, this divergence of interests in relation to energy co-operation is compounded by the commonly held view that, while the principles of the EU's market-based external and internal energy policies are well articulated, in practice a uniform degree of commitment from member states and institutions to these enunciated policies remains elusive. It is indeed no secret that a number of member states, including those with the bigger energy markets, have still to become reconciled to the Commission's belief that a consolidated and liberalized internal European energy market is in their national interests. The resistance that the Commission faced from member states like Germany, France and Italy in relation to its 2007 proposal for the 'unbundling' of the production and distribution segments of European energy incumbents is testimony to the inconsistencies that pervade European energy policy. Furthermore, from an SMC vantage point, the diluted unbundling model that the EU ended up adopting and the formalization of a requirement of reciprocity from third countries[21] that accompanied it is evidence that the 'EU is no less a geopolitical actor than other countries, but that the only difference is that its geopolitics are dressed up in the finer cloaks of rule-based discourse' (Youngs, 2009: 39).

The UfM, by contrast, seems to have eschewed the emphasis on rule-based co-operation with Mediterranean producers and has focused more on shared *practical* interests. This being said, the idea of developing North African solar power systems to supply Europe with renewable energy has been around for decades. What has allowed the MSP to gain unprecedented support on this occasion, presaging a more successful regional energy co-operation venture in the framework of the UfM, is the fact that it has been introduced at a time considered propitious by most, if not all, stakeholders. Indeed, the interest generated by the MSP, in the absence of almost any political and institutional progress on almost all other aspects of the UfM, suggests that the advances realized by the MSP so far owe little to the attributes of the EU's new Mediterranean policy. Rather, they seem to have benefited from the shifting energy interests of energy producers and consumers alike, pertaining more specifically to issues of climate change, energy security and growing energy requirements.

Coupled with concerns of energy security and external energy dependence, the commitments made by the EU in the context of the incipient international climate change agenda have led to a new focus of energy policy towards moving to a 'low-carbon' economy. In 2008, the EU detailed its plans in this regard under the overarching framework of '20–20–20': by the year 2020 reduce greenhouse emissions by 20 per cent; increase energy efficiency by 20 per cent; and raise to 20 per cent the share of renewable energy in its energy mix. As, in the meantime, solar technology is beginning to become more commercially viable for large-scale projects, the prospect of developing solar energy in the North African desert to meet these targets and at the same time diversify energy sources has become more attractive for politicians and industry decision makers alike.

This *recentrage* [recalibration] of EU energy policy priorities coincided with a growing realization on the part of SMC policy makers that not only was the depletion of their fossil fuel reserves no longer a distant prospect, but also that the domestic energy requirements of their own economies are growing at a phenomenal pace. Algeria, Egypt and Libya have all seen their domestic demand for gas and power grow at average annual rates of 8 per cent in recent years – a trend set to continue unabated in the coming decade. This has translated into tremendous pressure on their production and export capacity, setting in motion an urgent rethink of energy policies in SMCs. As a result, developing renewable sources of energy to satisfy this booming demand, thereby freeing more hydrocarbons for exports, and preparing the 'after-oil' era has gained more ground amongst SMC policy makers (Derradji, 2010). The consequence of these shifting priorities on both sides of the Mediter-ranean is that the MSP has rather suddenly appeared as a panacea to the region's corresponding energy and climate challenges, and is likely to yield better results as far as Euro-Mediterranean energy co-operation is concerned.

Thus, the outlook for energy co-operation under the UfM is to some extent a reflection of the renewed focus on functionalism that the new framework has (re-)introduced to Euro-Mediterranean relations. Its emphasis on the development of regional solar energy projects, rather than the neo-liberal, rule-based co-operation model its predecessor EU policies in the Mediterranean had pursued, is certainly finding resonance with SMCs, including petroleum-exporting countries. However, this owes more to the fact that renewable energy is currently receiving increasing attention within and beyond the region, as it is seen as a way of addressing a number of 'practical' energy and economic challenges that most interested parties are facing, than to the de-politicized focus of the UfM. In other words, what is encouraging Euro-Med partners to rally around the UfM solar project is a favourable market context, which even if the UfM had displayed a stronger high-politics emphasis would almost certainly have had the same effect on Euro-Mediterranean energy co-operation. In short, the UfM certainly represents some degree of change in Euro-Mediterranean relations, but crediting this change alone with reinvigorating energy co-operation in the region may be too stretched a conclusion, given that the relevant 'outside' context – in this case, energy market conditions – has known a more profound transformation compared to the days of the EMP and the ENP and is itself having a significant impact on the energy preferences of UfM partners and the consolidation of their co-operation.

**Conclusion**

This contribution has revisited the energy co-operation components of the EU's Mediterranean policy frameworks to date, namely the EMP, the ENP and the UfM. The aim was to compare the philosophies and approaches of these initiatives, with the aim of assessing the prospects of the UfM's regional energy co-operation agenda compared with the conspicuous failures of previous policy efforts. The argument is narrowly focused on the institutional and policy level of the EU's relations with petroleum-exporting SMCs, and as such excludes other analytical layers and actors, which in the field of energy are simply too complex and diverse to include in a relatively short contribution. Besides, the aim behind this parsimonious focus was to isolate the dynamics inherent in the EU's internal policy making as well as in its relations with SMCs in order to better assess the impact of their evolution on Euro-Med energy co-operation. As it happens, the findings of this study point to international energy market dynamics as being the most potent in influencing the prospects of the UfM in fostering regional energy co-operation – its functionalism only playing an intervening role.

What this contribution did not offer is a prognosis on plans to develop solar and renewable energy in the Mediterranean. What it did do, however, was highlight the impact that changing energy market conditions are having on the priorities of Euro-Med partners and how these, as opposed to the UfM's attributes, in turn will likely define the outlook for Euro-Med energy co-operation. Indeed, after 15 years of Euro-Mediterranean energy co-operation, the EU and its southern Mediterranean partners seem to have finally found common ground within the proposed energy co-operation model of the UfM. What the EMP and ENP failed to achieve in terms of interest convergence between European consumers and southern Mediterranean producer and transit countries, as well as in relation to the optimization of regional energy relations, the UfM's proposal for a solar plan seems to have realized quite effortlessly. However, rather than being reflective of some sort of natural selection process – i.e. the evolution of the EU's Euro-Mediterranean energy co-operation model on the back of a learning process – the resulting convergence of interests between energy consumers and producers in the Mediterranean is the product of processes that have occurred outside the space of interaction provided hitherto by the EU's policy frameworks.

## Acknowledgements

The author is grateful to Alfred Tovias, Amelia Hadfield, Gonzalo Escribano, Richard Gillespie, Federica Bicchi, Annette Jünemann and an anonymous reviewer for their valuable comments on an earlier draft of this study.

## Notes

[1] Fossil fuels (oil, gas and hard coal) account for 70 per cent of the EU's primary energy mix. In 2008, the EU imported over 50 per cent of its primary energy requirements – over 80 per cent of its oil supply and 60 per cent of its natural gas needs. By 2030, it is expected that it will import 65 per cent of its total energy requirements (Eurostat, 2010).

[2] Although the MSP aims to develop renewable energy not only from solar sources, as well as to promote energy efficiency, its solar component is likely to prove the most challenging and potentially most rewarding. This is why it is the main focus of this paper.

[3] Although it is possible that external (market) factors, such as oil prices, had also a role to play in the (under-)development of the EU's energy co-operation policies the Mediterranean, the analysis in this paper focuses primarily on variables/dynamics internal to the EU and to its relationship with SMCs.

[4] On these other aspects, see Hunt and Schlumberger (this collection).

[5] For Euro-Mediterranean development issues, see Hunt (this collection).

[6] These consisted of a meeting in Athens in July 1995 between the European Commission and the Mediterranean partners and a second meeting that took place in Madrid on 20 November 1995 between the energy ministers of Spain, France and Italy and representatives of SMCs.

[7] The Euro-Mediterranean Energy Forum has traditionally held its meetings at the level of directors-general. Its activities have been financed by Euro-Med instruments MEDA and ENPI.

[8] Recent projects include the Euro-Arab Mashreq Gas Market, Maghreb Electricity Market Integration, MED-ENEC (a project promoting energy efficiency in the construction sector in the Mediterranean), and MED-REG (a project of association of Mediterranean energy regulators). For more details, see European Commission (2008a).

9   It is worth pointing out that Libya has never participated as a full member in the EMEP because it has refused to adhere to the EMP's *acquis*.
10  The state-owned gas companies in exporting SMCs saw in the liberalization of the European gas market a threat to their commercial interests because of the impact this could have on their existing margins, market shares and the future of long-term contracts which had traditionally spread the risk of capital-intensive gas projects fairly between producers and consumers through price and volume mechanisms.
11  Destination clauses, or territorial restrictions, were a standard feature of long-term supply contracts in Europe for both pipeline gas and LNG. They seek to restrict the ability of the buyer to on-sell gas to a third party outside its territory at a higher price.
12  See 'Memorandum of Understanding on Strategic Partnership on Energy between the European Union and the Arab Republic of Egypt', available at: http://ec.europa.eu/external_relations/egypt/docs/mou_energy_eu-egypt_en.pdf (accessed 10 January 2010).
13  See 'Spanish gas groups fear reliance on Algeria', *Financial Times*, 12 May 2008; 'Chakib Khelil: Sonatrach dérange', *Le Quotidien d'Oran*, 12 March 2009.
14  See 'Sarkozy seeks to pair French, Algerian natural-gas firms', *The Wall Street Journal*, 29 June 2007.
15  Existing contracts were due to expire in 2013.
16  The other five projects are: 1) de-pollution of the Mediterranean; 2) maritime and land highways; 3) civil protection; 4) higher education and research; and 5) small and medium-sized business development. For more details see: Joint Declaration of the Paris Summit for the Mediterranean, 13 July 2008, available at: http://www.ue2008.fr/webdav/site/PFUE/shared/import/07/0713_declaration_de_paris/Joint_declaration_of_the_Paris_summit_for_the_Mediterranean-EN.pdf (accessed 10 January 2010). Final Statement of UfM foreign ministers meeting in Marseille, 3–4 November 2008, available at: http://ue2008.fr/webdav/site/PFUE/shared/import/1103_ministerielle_Euromed/Final_Statement_Mediterranean_Union_EN.pdf (accessed 10 January 2010).
17  See the 'Mediterranean Solar Plan Strategy Paper', available at: http://ec.europa.eu/energy/international/international_cooperation/doc/2010_02_10_mediterranean_solar_plan_strategy_paper.pdf (accessed 10 January 2010).
18  Another important aspect of the MSP consists of encouraging energy efficiency in the region, considering the relatively high energy intensity rates in the MENA region. For more details, see The World Bank (2009).
19  Since then, a number of SMC-based companies have subscribed to the initiative, such as Algeria's private operator Cevital. See for more details: http://www.desertec.org
20  It is beyond the scope of this paper to delve deeper into these challenges, but for more details see: PriceWaterhouseCoopers (2010), RaL (2010), Laffitte et al. (2009) and German Aerospace Center (2006).
21  This provision is now infamously known as the 'Gazprom clause', as it was perceived as being aimed at limiting the inroads that the Russian gas giant has been making into the European energy market. Its aim is said to be to attach (SMC) national oil companies' access to the European gas (and electricity) downstream markets to conditions pertaining to the terms imposed by their governments to European investment in the upstream segments of their hydrocarbon markets.

# References

Aïssaoui, A. (2002) European strategy for the security of energy supply: re-evaluating relations between the EU and the producers and transit countries of North Africa, *Middle East Economic Survey*, 45(15), pp. D1–D9.

*Barcelona Declaration* (2005) Barcelona declaration adopted at the Euro-Mediterranean Conference 27-28/11/95. Available at http://trade.ec.europa.eu/doclib/docs/2005/july/tradoc_124236.pdf (accessed 10 January 2010).

Council of the European Union (2000) Common strategy of the European Council of 19 June 2000 on the Mediterranean Region, 2000/458/CFSP, *Official Journal of the European Communities*, 22 July, L 183/5–10.

Council of the European Union (2003) Ministerial declaration of the Euro-Mediterranean Energy Forum, 21 May, Athens.

Council of the European Union (2004) Final report: EU strategic partnership with the Mediterranean and Middle East, Brussels, 18 June.

Council of the European Union (2007) Ministerial declaration on the Euro-Mediterranean energy partnership, 1607/07, 17 December, Limassol.

Darbouche, H. (2007) Russian–Algerian cooperation and the 'gas OPEC': what's in the pipeline?, CEPS Policy Brief No. 123, Brussels.

Derradji, B. (2010) Developing renewables in Algeria: the case for CSP, presented at the Valencia conference on the Mediterranean Solar Plan, 11 May.

Escribano, G. (2010) Convergence towards differentiation: the case of Mediterranean energy corridors, *Mediterranean Politics*, 15(2), pp. 211–229.

European Commission (1994) Strengthening the Mediterranean policy of the European Union: establishing a Euro-Mediterranean partnership, COM (94) 427 final, Brussels.

European Commission (1995a) Green paper: for a European Union energy policy, COM (94) 659 final 2, Brussels.

European Commission (1995b) White paper: an energy policy for the European Union, COM (95) 682 final, Brussels.

European Commission (1995c) Strengthening the Mediterranean policy of the European Union: proposals for implementing a Euro-Mediterranean Partnership, COM (95) 72 final, Brussels.

European Commission (1996) Communication from the Commission to the European Parliament and Council concerning the Euro-Mediterranean Partnership in the energy sector, COM (96) 149 final, Brussels.

European Commission (2000a) Green paper: towards a European strategy for the security of energy supply, COM (2000) 769 final, Brussels.

European Commission (2000b) Reinvigorating the Barcelona Process, COM (2000) 497 final, Brussels.

European Commission (2001) Enhancing Euro-Mediterranean cooperation on transport and energy, COM (2001) 126 final, Brussels.

European Commission (2003a) Wider Europe – neighbourhood: a new framework for relations with our eastern and southern neighbours, COM (2003) 104 final, Brussels.

European Commission (2003b) Communication from the Commission to the European Parliament and the Council on the development of energy policy for the enlarged European Union, its neighbours and partner countries, COM (262) 262 final/2, Brussels.

European Commission (2004) European Neighbourhood Policy: strategy paper, COM (2004) 373 final, Brussels.

European Commission (2006a) Press release: energy commissioner Piebalgs welcomes the Euro-Mediterranean energy forum that takes place today in Brussels, IP/06/1238, 21 September, Brussels.

European Commission (2006b) An external policy to serve Europe's energy interests: paper from the Commission/SG/HR to the European Council, S160/06, Brussels.

European Commission (2006c) Communication from the Commission to the European Parliament and the Council on strengthening the European neighbourhood policy, COM (2006) 726 final, Brussels.

European Commission (2006d) Green paper: a European strategy for sustainable, secure and competitive energy, COM (2006) 105 final, Brussels.

European Commission (2007) An energy policy for Europe, Brussels, COM (2007) 1 final, Brussels.

European Commission (2008a) EMP regional cooperation: an overview of programmes and projects, Brussels.

European Commission (2008b) An EU energy security and solidarity action plan, COM (2008) 781 final, Brussels.

Eurostat (2009) Panorama of energy: energy statistics to support policies and solutions, European Commission, Brussels

Ferrero-Waldner, B. (2009a) The Mediterranean Solar Plan – a necessity, not an option, speech at the EU Sustainable Energy Week, 13 February, Brussels.

Ferrero-Waldner, B. (2009b) Energising the Euro-Mediterranean region: the Mediterranean Solar Plan, speech at the Third European Renewable Energy Policy Conference, 17 November, Brussels.

German Aerospace Centre (2006) Trans-Mediterranean Interconnection for Concentrating Solar Power – Final Report, Stuttgart. Available at http://www.dlr.de/tt/Portaldata/41/Resources/dokumente/institut/system/projects/TRANS-CSP_Full_Report_Final.pdf

Hallouche, H. (2006) The Gas Exporting Countries Forum: is it really a gas OPEC in the making?, Working Paper No. 13, Oxford Institute for Energy Studies, June.

Laffitte, M. et al. (2009) 'Rapport sur le Plan solaire méditerranéen', Paris. Available at http://lesrapports.ladocumentationfrancaise.fr/BRP/094000284/0000.pdf

PriceWaterhouseCoopers (2010) 100% Renewable Electricity: A Roadmap to 2050 for Europe and North Africa, London. Available at http://www.pwc.co.uk/pdf/100_percent_renewable_electricity.pdf

Resources and Logistics (RaL) (2010) Identification Mission for the Mediterranean Solar Plan – Final Report, Brussels. Available at http://ec.europa.eu/energy/international/international_cooperation/doc/2010_01_solar_plan_report.pdf

World Bank (2009) Tapping a hidden resource: energy efficiency in the Middle East and North Africa, Energy Sector Management Assistance Program (ESMAP), Washington DC.

Youngs, R. (2007) Europe's external energy policy: between geopolitics and the market, CEPS Working Document No. 278, Brussels.

Youngs, R. (2009) *Energy Security: Europe's New Foreign Policy Challenge* (London: Routledge).

# The UfM Found Wanting: European Responses to the Challenge of Regime Change in the Mediterranean

## RICHARD GILLESPIE

*This final contribution takes analysis of the Union for the Mediterranean into its third year by focusing on how European actors attempted to respond to the wave of popular insurrections in the Arab world during the early months of 2011. While events on the ground have opened a new window of opportunity for Europe to exert renewed influence in its southern neighbourhood by focusing activity around support for democratic transformations, the UfM has been largely by-passed in European responses to the Arab Spring. While this Union continues to struggle to operate at a modest level, European political action is being channeled largely through a revised European Neighbourhood Policy. Apart from the design defects that have been present from the start, the UfM's shortcomings need to be seen as symptomatic of the crisis surrounding the European project: the problematic intergovernmental structure is in evident need of reform, yet policy renewal is hampered, among other things, by wider European controversies over implementation of the Lisbon Treaty.*

The sense of failure surrounding the Union for the Mediterranean persisted throughout 2010, only for its limitations to be brought into even sharper relief the following year when faced with the new challenge posed to Europe policy-makers by the triumphs and frustrations of pro-democracy movements in the Arab world. A further attempt to hold a second UfM summit failed in November 2010, again owing to Arab disappointment at the unproductive outcome of US efforts to revive Middle East peace talks. Two months later, further symptoms of crisis were evident when the UfM secretary-general, Ahmad Masa'deh, resigned, acknowledging that the institution he had headed had been unable to launch the planned new Euro-Mediterranean projects, partly owing to insufficient finance.[1] Also in February, Italian foreign minister Franco Frattini described the UfM as an 'empty shell'. It was against this backcloth of unproductive drift in the Union for the Mediterranean that mass movements emerged to dispute the power of authoritarian Arab regimes, some of which were prominent partners of the EU in the Euro-Mediterranean framework. Indeed, at the time of his downfall, former Egyptian ruler Hosni Mubarak was still co-President of the UfM.

Particularly embarrassing for President Sarkozy, and forcing him to replace foreign minister Michèle Alliot-Marie, was the triumph in mid-January 2011 of the 'revolution of dignity' in Tunisia. There the involvement of France with the regime of Azzerdine Zine El Abidine Ben Ali had gone far beyond *realpolitik* to include the personal business interests and luxury holidays of ministers. Yet what brought the UfM itself more clearly into disrepute was the ensuing uprising in Egypt, leading to the resignation of Mubarak in February: the massive popular repudiation within his own country of the man who had been chosen by Sarkozy to be co-President of the UfM drew international attention to the politically unscrupulous nature of the Union for the Mediterranean and the deeply unsatisfactory nature of its highest level of governance. In general, the uprisings in southern Mediterranean countries brought into question the European political perspectives that had informed the Barcelona Process and especially the neofunctionalist vision of the UfM, both of which had played down 'growing instability at the local level and protest against regimes in North Africa' (Kausch and Youngs, 2009: 775). The European Commissioner for Enlargement and Neighbourhood Policy, Stefan Füle, candidly commented that EU officials needed to acknowledge the mistake of thinking that democracy was not viable in the Arab world (*El País*, Madrid, 19 March 2011).

Yet, although some voices now proclaimed the UfM to be a failure and called for it to be abolished,[2] this was no means the mainstream reading of the significance of the 'Arab Dawn' made by European political leaders and EU officials, concerning the future of this Euro-Mediterranean structure. Rather, as will be outlined below, there were a number of compelling reasons for *not* replacing the UfM. Most of these related to the general situation prevailing in the EU as actors responded to the implementation challenges of the Lisbon Treaty and sought to define its actual significance for the future of multi-level governance of Europe.

## Not a Good Time for a Radical Rethink?

With mainstream European opinion welcoming the popular struggles to overthrow authoritarian regimes in the Middle East and North Africa, European governments and the EU faced public pressure to devise concrete means of supporting pro-democracy movements and counteracting violently repressive responses to them from a number of rulers. The Union for the Mediterranean, with its emphasis on technical cooperation, was by no means an adequate vehicle (even if it could gain momentum) to deliver support to democratizing efforts in the South; yet what seemed to some to be a window of opportunity to renew EU Mediterranean Policy, comparable to the early 1970s and early 1990s, did not in fact give rise to a radical rethink. Instead, although European policy-makers called for a 'reform' or 'refounding' of the UfM, what they meant by this was merely a refocusing of policy emphasis and renewed effort to make at least some of its structures operational. In particular, it was hoped to replace Masa'deh with a more pragmatic secretary-general, who might make a difference to the capacity of the Secretariat to launch projects. The only half-criticism of the UfM expressed in the first policy document from Brussels following the Arab uprisings was that 'its implementation did not deliver the results we expected. The UfM needs to reform to fully realize its potential' (European Commission/High Representative, 2011a: 11).

A strategic rethink leading to a new approach to Euro-Mediterranean relations held little appeal within policy circles during the early months of 2011, for a number of reasons. First and foremost was the fact that events moved very quickly, both within some Arab countries and across the region, prompting demands within Europe and among pro-democracy Arab activists that international organizations such as the EU act with urgency to assist mass movements on the ground, struggling for regime change, and to mobilize financial resources to support stabilization efforts once new authorities emerged. Rather than engage in a strategic policy review, the prevailing attitude in the European Union was that there was simply no time: urgency was of the essence, thus it was far better to revisit the existing policy wardrobe and find items there that could be put to use more or less immediately, after minor adjustments. Given the multifarious nature of the preceding Barcelona Process, there was no shortage of old proposals and discontinued lines that could be put together in a new policy offer, aimed at supporting democratization and political reform where local conditions now provided more propitious circumstances to advance in these directions. In fact, despite much effort on the part of certain officials and MEPs, the EU proved too cumbersome to deliver a dynamic response.[3]

Moreover, in the case of the European Neighbourhood Policy, a policy review had already been initiated way back in July 2010, months before the Arab upheavals commenced.[4] With the review process, as initially conceived, coming close to completion, there was very little support in Brussels for having a complete rethink in response to the fact that Arab people in huge numbers were demanding and risking their lives for a democratic future, providing evidence that the old dictatorships could in fact be brought down by popular insurrections. Instead, the existing policy review was extended so that adjustments could be made to the outcome. Essentially, the review brought proposals to make democratizing reforms more central to the policy content of future ENP Action Plans, while reinforcing political conditionality and modifying the inter-governmental nature of the framework in order to facilitate cooperation at the level of civil society.[5]

Second, there was recognition within the EU that the 'danger' perceived in 2007-8 of President Sarkozy's initiative resulting in a French-led 'regionalization' of (European) Mediterranean policy had been seen off by this time. For many critics and sceptics, the UfM's early intergovernmentalism had begun to be moderated by the end of the decade, partly as a result of the successive replacements of those foreign ministers who had been champions of the UfM initiative. Former ministers Bernard Kouchner in France and Miguel Ángel Moratinos in Spain were are no longer involved and their successors, Alain Juppé (following the brief spell of Michèle Alliot-Marie in the Quai d'Orsay) and Trinidad Jiménez, brought greater acceptance of an enhanced EU role.[6] By April 2011, both countries had made clear that they no longer aspired to future roles in the European component of the UfM co-Presidency (although other member states had not exactly been encouraging them to pursue such ambitions in the first place). This represented an acceptance of EU coordination within the UfM, with central actors (the High Representative, supported by the EEAS, and the European Commission) providing the European input into governance structures, rather than this being delegated to individual Mediterranean member states. Moreover, by this time, France was desperately trying to make amends for its traditional support for authoritarian Arab regimes by advocating the adoption of new UfM projects with some kind of civil society focus and a greater effort around democracy promotion and support through the European Neighbourhood Policy.[7]

Meanwhile, there was no desire among other EU member states and Mediterranean partners to humiliate the French progenitor of the UfM, already embarrassed by the revelations about French links to the Ben Ali regime and popular repudiation of Mubarak in Egypt. Respect for Sarkozy's sensitivities, whether because France now appeared to be more accommodating or because of independent reasons for countries to want good relations with Paris, was widespread, to the extent that actual replacement of the UfM was not contemplated, 'at least not before a change of President in France'.[8] Tension between France and other EU actors over Mediterranean issues did not disappear entirely, however, and there were fresh mutterings about Sarkozy's political style, prompted by his *coup de main* in March, when he attempted to put pressure on colleagues to recognize the Libyan Transitional National Council: France's own recognition was announced the day before the extraordinary meeting of the European Council to discuss EU responses to the Libyan conflict and the Arab Spring in general, in Brussels on 11 March 2011.

A further impediment to a profound process of policy renewal was the lack of a national protagonist or policy entrepreneur with a new vision for EU Mediterranean policy. While France and Spain had dominated earlier renewal efforts and now drew back, Italy had never been more than a secondary player, and now seemed single-mindedly preoccupied with the migratory effects of the turmoil in Tunisia and conflict in Libya. Other southern European member states, Greece and Portugal, were (like Spain) preoccupied with financial crises. In Spain, Zapatero drew criticism for a belated response to the Arab Spring and for double standards in his dealings with North African and Gulf states when he did try to reposition his country as 'a cheerleader for the process of reform', while falling in with the view that

Tunisia should be prioritized (Echagüe, 2011: 2; Echagüe et al., 2011 forthcoming).[9]

Finally, the UfM appeared to be condemned to a prolongation of its life—even if devoid of vivacity—by controversy within Europe over the application of the Lisbon Treaty, affecting the EU's governance structures. While there was still little enthusiasm for the UfM within EU institutions, those actors that were still interested in getting the UfM off the ground looked chiefly to an enhanced EU role within it to provide momentum. Yet it was precisely this 'solution' that alarmed certain member states, concerned that an unacceptable precedent might be set, with consequences far beyond the realms of Mediterranean policy. What was done to 'reform' the UfM in the wake of the Arab uprisings could help define the meaning of the Lisbon Treaty, with lasting effects upon the power configuration within the EU.

In response to the initial set of policy proposals for a 'Partnership for Democracy and Shared Prosperity with the Southern Mediterranean', the UK in particular expressed objections, independently of substantive policy debates, because of wider concerns about EU 'competence creep'. The British government was alarmed by the fact that a document co-authored by the European Commission and the High Representative of the EU for Foreign Affairs and Security Policy stated that both were 'ready to play a bigger role in the Union for the Mediterranean in line with the Lisbon Treaty' (European Commission/High Representative, 2011b: 11). The inter-governmentalist approach associated with the UfM was quite acceptable to the Cameron government, because it allowed the UK to block or opt out of any Euro-Med activities that it opposed, or that might set a precedent for an enhancement of supranationalism. The UK took the extreme line that the UfM had nothing to do with the EU, whereas most other member states were pleased to see it becoming more Europeanised and were willing to let the Secretariat play a role in relation to sub-regional projects (regional ones being unlikely owing to the Middle East conflict).[10]

There was still, however, a clear possibility that central actors would take over the EU's involvement in the co-Presidency of the UfM if, in order to placate those concerned about Lisbon, the European Council were to issue an explicit mandate for this to happen; but clearly the procedural wrangling meant that there would be no quick and easy solutions to the problems afflicting Euro-Mediterranean co-governance. Meanwhile, the need to prepare for and hold competitive elections in Tunisia and Egypt meant that time was required before fresh representation could emerge among Mediterranean partners, not least because Tunisia was seen as a possible candidate to serve as the next non-European co-President. Before this, however, the new Tunisian authorities had to be elected (in October, if all went well) and even then the Arab group would need to reach a consensus over the issue.

## Announcing a New Partnership...through the ENP?

Thus, 'reform' of the UfM became mainstream thinking in the EU, but what it actually meant was a degree of pragmatic compromise over rather modest efforts to make this framework operational, at least at some levels. In fact, as the 'Partnership for Democracy and Shared Prosperity' proposal implied as early as March 2011, the EU response to political change in the Arab world would be managed largely through the ENP, identified thus far with individual bilateral dealings with neighbours. The March document made only brief and vague references to the UfM while its emphasis on more 'differentiation' in relations with Arab partners ('those that go further and faster with reforms will be able to count on greater support from the EU') implicitly envisaged the Neighbourhood policy as the main mechanism for providing reform incentives (European Commission/High Representative, 2011a: 11, 5).[11]

On how to make the UfM functional, many looked to aspects of the Lisbon Treaty, the appointment of a new secretary-general of the Secretariat and the eventual emergence of new leaders and representatives in some of the Arab countries, shifting the political balance sufficiently for the co-Presidency to be made to work. However, whether there would be less of the disruption to activity caused by the vulnerability of co-governance structures to fallout from the Arab-Israeli conflict was by no means certain. While it was expected that Arab governments emerging from free, competitive elections would tend to be keen to work with the EU around a wider range of policy issues, including democratization agendas, it was also anticipated that they would strive to convince their electorates that they were taking principled positions in support of the Palestinians. Mainstreaming of the Palestinian cause by the Arab group could be expected to continue to the extent that the status quo prevailed in hard-line states such as Syria, though certainly the group stood to become politically more diverse and unity more difficult to maintain as reform incentives increased.

In the immediate aftermath of the downfall of rulers in Tunisia and Egypt, it still did not prove possible for European representatives to secure agreement with the Arab group (still coordinated by Egypt) on the holding of a UfM summit or even a foreign minister's conference. As in the recent past, this was partly because of what such events meant for the Arab partners in terms of relations with Israel. Equally, with some Arab rulers clinging to power through a massive use of repression, there was now less interest among European leaders in being seen rubbing shoulders with them at Euro-Mediterranean meetings.[12] Initially, at least, the continuing impossibility of holding high-level meetings meant that even modest reform of the UfM was frustrated, simply because this left no other mechanisms in place to undertake renewal of the co-Presidency or to approve new projects.

In the early months of the Arab Spring, UfM meetings continued to achieve little since they were feasible on a regular basis only at senior official level, where representatives felt they lacked the authority to reform or innovate through reaching decisions on new project areas.[13] The most that had been achieved by mid-year was a call for new project proposals in the fields of transport and urban development, for which the Secretariat would establish working groups in the coming months.[14] This suggested that UfM project activity would continue to be 'technical' rather than politically-focused.

At the same time, initiatives were held up by Europeans debates over what the EU's priorities should be and especially the extent to which concessions might be made to meet Arab requests over issues such as visa liberalisation and greater market access for southern agricultural goods, where European security outlooks and economic interests had traditionally upheld the status quo. It is noteworthy that the declaration that emanated from the extraordinary European Council on 11 March, while 'broadly' welcoming the communication from the Commission and High Representative, put much more emphasis on European security concerns, notably those associated with migration (European Council, 2011).

Overall, there seemed a better prospect of a European consensus emerging over future action through the ENP for, through this framework, improved offerings to southern neighbours could be negotiated bilaterally and not with Mediterranean partners or the Arab group as a bloc. At a time of financial stringency, and with European leaders trying to lower any expectations of a 'Marshall Plan' in response to the Arab Spring, increased differentiation in dealing with Arab partners not only made sense in terms of political values, but would involve less sacrifices by Europeans owing to provisions for enhanced conditionality, depending on the extent to which neighbours reformed their systems. Conditionality would now be emphasized much more than in the early years of the ENP. Then, countries like Ben Ali's Tunisia could keep democracy and human rights off the agenda by showing enthusiasm for reform in the socio-economic domain. Now, commitment to democratisation would be mainstreamed and the vision of ENP officials involved an offer of 'more for more'—greater benefits for countries that decided to converge with Europe *across the board*.

The effectiveness of conditionality will depend on whether the EU can agree upon and actually deliver enhanced benefits through new ENP Action Plans. Early divergence among European actors was expressed over some key issues. Mention has been made already of the resistance from some member states to the March proposals in areas such as market access and human mobility, but there may at least be scope for negotiation in relation to the longstanding issue of balance between financial aid and market access.[15] Increased funding, to come chiefly from the European Investment Bank and European Bank for Reconstruction and Development, does not itself seem to be a major issue and 'could make a difference' (Emerson, 2011), yet there remains some disagreement over how conditionality should operate. While the UK is among those that want to see prior approval of reforms and the actual fulfillment of reform plans to condition the granting of funds, others maintain that increased aid needs to accompany reforms if they are to succeed. While stricter conditionality requirements are seen by critics as being in some cases a cover for parsimony, more flexible approaches are seen by other critics as a continuation of traditional ENP practices that cannot claim a strong record of success. Southern European countries still tend to stick to advocacy of rewarding reformists but not punishing non-reformers (*El País*, 10 March 2011), but following the last round of EU enlargement they are now clearly in a minority.

The ENP has the great advantage over the UfM that its bilateral structure ensures that, while a specific set of relations may be disrupted by disputes, this does not affect the other relationships; for all can develop at variable speeds. However, the ENP does have limitations and should not be seen as a *sufficient* basis for a renewed partnership. For one, it does not involve all the EU's southern neighbours. Libya, one of the biggest challenges for the EU, remains outside; and while Algeria qualifies for participation, it has yet to show any interest in an agreement beyond the energy field. Second, for the EU to put its whole emphasis on the ENP would mean to give up on Euro-Mediterranean multilateralism completely, notwithstanding the fact that—in the view of many, and despite widely recognized shortcomings—the Barcelona Process can be credited with some achievements, such as the creation of the Anna Lindh Foundation. Enhanced differentiation and conditionality could help accentuate divisions among Mediterranean partners and bring a further retreat from region-building. This may be a matter of the EU responding to situations on the ground, where some authoritarian regimes have proved more resistant than others, but there is also some element of the Union taking the easier, less risky option and acting without a long-range strategy.

## Bilateralism in need of Regional Complementation?

While reform or replacement of the UfM are unlikely to happen any time soon—and may have to wait until new governments emerge in a number of countries—the regional level of partnership is certainly one that requires further thought and initiative if the EU wishes to play a positive role in facilitating improved horizontal relations among Mediterranean neighbours. Looking forward, strategic thinking about Euro-Mediterranean policy frameworks will need to dwell on two concepts, provided for under the UfM but not yet finding expression in practical ways of structuring relationships in a functional manner. One of these is 'variable geometry'. Three years after the launch of the UfM, there was still no agreement about the rules under which countries could collaborate in, or deny other partners involvement in, UfM projects: whether these should be open to all members or there should be some provision allowing a founding group to exclude others whose involvement was deemed likely to prove disruptive. The other is co-ownership. This has proved problematic in the UfM, as a result of being structured around southern European and Middle Eastern representation. The bipolar approach, whose origins lie in the simplistic notion that Euro-Mediterranean relations are just a North-South form of partnership, has had the effect of excluding northern and eastern European member states and western Balkan EU accession countries from the governance structures of the UfM, leaving many participants feeling alienated. Before the Arab Spring, there was also a feeling among countries of the Maghreb that the Egyptian co-Presidency had shown little interest even in representing the Arab group as a whole.

It may be that the space covered by the UfM is simply too heterogeneous for the concept of co-ownership to be applied, at least at the macro-level extending to 43 countries. At the same time, it would not be easy to downsize, its member countries having formally signed up to it at the highest political level just a few years ago. There is still a case for the UfM collectively to acquire value as a forum for discussion and cooperation in policy areas such as security that have a true Mediterranean dimension (Hanelt and Möller, 2011: 6). For actual cooperation to take place, however, minilateral constructs might prove more effective than multilateral ones, and if a root and branch reconsideration of Euro-Mediterranean relations lacks support at the present time, then some form of 'tiering' of Euro-Mediterranean relationships under the umbrella of the UfM would seem to be what is needed to allow 'sub-regional' forms of cooperation to be established, based on groupings of countries that have a consistent common interest in engaging in long-term dialogue and activity in a particular field, and are held back at present by the disinterest or political hostility of other UfM countries.

'Sub-regional' cooperation has tended to be considered in the past primarily in terms of lesser geographical spaces such as the western Mediterranean, deemed to be more coherent or less affected by conflict, yet one consequence of the Arab Spring may be to throw up new international coalitions defined largely in political terms, such as those that want to see regime change, albeit with representative structures assuming a variety of political forms. Thus a sub-regional grouping could take the form of EU countries keen to promote and support democratization, or even EU states as a whole, in conjunction with partners that show real commitment to a similar agenda, plus the Commission and EEAS. This kind of approach was hinted at in the 'Partnership for Democracy and Shared Prosperity' document, which proposed that a commitment to 'adequately monitored free and fair elections' should be 'the entry qualification for the Partnership' (European Commission/High Representative, 2011a: 5).

However, that document did not advocate any institutionalisation of the proposed partnership. The institutional setting remained dominated by the ENP and the much more problematic UfM. The subsequent Neighbourhood document did refer vaguely to a new aim to 'advance sub-regional cooperation' and to a regional dimension being added to the ENP, while its specific references to the UfM mostly reiterated the latter's founding statement of intent. The ENP document affirmed that 'sub-regional co-operation involving fewer neighbours and concentrating on specific subjects can bring benefits and can create greater solidarity', yet appeared to evade the strategic challenge of engaging in democratic political solidarity by giving as an example action in support of greater physical interconnectivity in the Maghreb (European Commission/High Representative, 2011b: 15, 17). The review of the ENP supported proposals to establish a European Endowment for Democracy, as proposed by Poland, and a Civil Society Facility for the Neighbourhood (European Commission and High Representative, 2011b: 4), but these interesting ideas sounded decidedly European-centred, albeit with regional reach, rather than opportunities in which co-ownership might be encouraged.

## Conclusion

Faced with the political upheavals in neighbouring Arab countries, the shortcomings of the Union for the Mediterranean were underlined both by the persistence of very limited forms of activity and by the very modest, almost perfunctory mentions of it in European policy documents issued to outline proposed EU responses to the events of the Arab Spring. For all intents and purposes, the UfM had been at an impasse at the political level ever since the Marseille foreign ministers meeting in late 2008. And despite the arrival of the Arab Spring, it has continued to receive scant attention at senior levels within EU institutions, at a time when the map of the Mediterranean is much more prominent in the minds of European politicians and officials than it was earlier. Even if it proceeds to launch projects in line with the original French intent, it now seems destined to be overshadowed by the ENP as a channel for European relations with Mediterranean partners and for responses to the Arab Spring.

The UfM is being allowed a further opportunity to show what it can do, primarily because of the political capital invested in it by France, and the involvement of Spain through the existence of the Secretariat in Barcelona. Equally, the fact that it has been retained, albeit in a rather limited form, would seem to reflect a vague lingering recognition that something additional to the European Neighbourhood approach is needed, with a regional focus, even if the ENP itself can be enhanced.

Apart from the design defects that have plagued the Union for the Mediterranean from the beginning, it has suffered from an EU context in which belief in further European integration has been questioned quite radically, not least at the level of external relations, as decision making on major issues has tended to migrate back to the capitals of the larger member states. The UfM has been symptomatic of this trend, yet does not simply reflect it: its own development has been severely hampered, not only by spillover from the Middle East conflict, but by its own internalized intergovernmentalism, owing to the friction that this has engendered or exacerbated among European political actors. It has been found wanting at a time when a window of opportunity has opened for greater European engagement with Arab neighbours and Mediterranean policy renewal, yet the European response has been constrained by the need for urgency, various vested interests in existing institutional structures and the absence of a convincing policy entrepreneur, especially among the member states. Meanwhile some member states have resisted the efforts of central EU actors to provide a lead in response to the need for more political impetus and to meet the demands for support coming from Arab reformists.

Unlike the Barcelona Process, which generated early enthusiasm and then had its ups and downs before eventually declining into lethargy, the UfM experienced prolonged crisis almost from day one and is still operating at a low level. None the less, it seems destined to persist with its efforts to make a difference to Euro-Mediterranean relations, not least because the preconditions for a more ambitious form of region-building simply do not exist at the present time.

## Acknowledgements

Research for this chapter included a number of interviews with EU officials from the European Commission and European External Action Service, conducted in March 2011. It was made possible by an award from the British Academy for a project on 'The Union for the Mediterranean: Significance for the Barcelona Process' (SG-51979). The author is grateful for feedback on an earlier draft from Federica Bicchi.

## References

Echagüe, A. (2011) 'Time for Spain to Lead the EU's Mediterranean Policy', FRIDE Policy Brief 74, Madrid: Fundación par alas Relaciones Internacionales y el Diálogo Exterior, April.

Echagüe, A., Michou, H. and B. Mikael (2011) 'Europe and the Arab Uprisings: EU Vision Versus Member States Action', *Mediterranean Politics* 16/2, forthcoming.

Emerson, M. (2011) 'Review of the Review – of the European Neighbourhood Policy', Centre for European Policy Studies, Brussels, *European Neighbourhood Watch* 71, May.

European Commission and High Representative of the Union for Foreign Affairs and Security Policy (2011a) 'Joint Communication to the European Council, the European Parliament, the Council, the European Economic and Social Committee and the Committee of the Regions, *A Partnership for Democracy and Shared Prosperity with the Southern Mediterranean*', Brussels, 8 March, COM(2011) 200 final.

European Commission and High Representative of the Union for Foreign Affairs and Security Policy (2011b) 'Joint Communication to the European Parliament, the Council, the European Economic and Social Committee and the Committee of the

Regions, *A New Response to a Changing Neighbourhood*, Brussels, 25 May, COM (2011) 303.

European Council (2011) 'Cover Note from General Secretariat of the Council to Delegations, Extraordinary European Council, 11 March 2011, *Declaration*', EUCO 7/11.

Hanelt, C.P. and A. Möller (2011) 'How the European Union Can Support Change in North Africa', Bertelsmann Stiftung, *Spotlight Europe*, 2011/01, February.

Kausch, K. and R. Youngs (2009) 'The End of the Euro-Mediterranean Vision', *International Affairs* 85/5, pp.963-75.

## Notes

1 As with the inability to hold summit meetings, the UfM subsequently proved over-ambitious with regard to the timetable agreed for replacing Masa'deh and the original deadline for applications had to be put back to allow more time for candidates to emerge. A single candidate eventually emerged and was appointed in May, career diplomat Youssef Amrani, former secretary-general of the Moroccan Ministry of Foreign Affairs (*Jeune Afrique*, 18 May 2011).
2 For example, the *Financial Times*, 10 March 2011, described the UfM as 'a wretched foreign policy instrument if ever there was one' and said it should be wound up.
3 Jordi Vaquer, 'Europa dimitida', *El País*, 10 May 2011, described the response as 'bureaucratic' and cited the length of time taken over reform of the European Neighbourhood Policy.
4 Interviews with John O'Rourke, head of the European Neighbourhood directorate and Tomás Duplá del Moral, head of the Middle East/Southern Mediterranean directorate, European External Action Service, March 2011.
5 It was finally published on 25 May 2011.
6 Sarkozy's foreign policy, including his choice of Ben Ali and Mubarak as partners in the UfM and a failure to listen to his ambassadors in North Africa, was publicly criticized by a group of French diplomats in *Le Monde*, 28 February 2011.
7 Even now, however, when trying to justify the French shift from supporting authoritarian regimes to backing elected governments, Sarkozy's televised argument was still about security: stable democracies were needed to avoid uncontrolled migration to Europe (*El País*, 28 February 2011).
8 Off-the-record interview with an EU official, March 2011.
9 The argument here was that EU resources and effort should be concentrated on Tunisia owing to its relatively small size and population, recent experience of sustained growth, advances in the position of women in society, the pacific nature of its Islamist movement, the country's experience of cooperation with the EU and so on, all of which made it one of the Arab countries with the best prospects of successful democratization. If the EU could play an important support role, then Tunisia could become a 'showcase', encouraging efforts to achieve 'deep' democratization elsewhere. For some European actors, this made good strategic sense, while in the case of others, one suspects that the concentration on Tunisia was being advocated simply with the objective of ensuring that the EU and its member states did not become too involved in domestic power struggles as the movement for regime change spread to more and more Arab countries.
10 Interview with an EU official, March 2011.
11 One EEAS official (interview, March 2011) described the March document as 'ENP in substance' and said the 'meat' of EU Mediterranean policy lay in bilateral relations.

12 'In Need of a Lifeline', *Financial Times* feature, 19 April 2011.
13 Events that did prove possible were a UfM Assembly on 4 March 2011, at which parliamentarians exchanged ideas on future activity, and an eighth Euro-Mediterranean meeting on Industrial Cooperation, in Malta on 11-12 May. Initiatives to hold other ministerial conferences were blocked by Syrian insistence that they should have a Middle Eastern focus.
14 'UfM ready to receive urban development and transport project proposals', ENPI Information Centre, at http://www.enpi-info.eu/?lang_id=450, downloaded 31 May 2011.
15 A French diplomat, interviewed in March 2011, indicated that he expected France to be more flexible on market access for southern agricultural produce so long as it could obtain a balancing commitment from other member states through the financial package.

# Index

Page numbers in *Italics* represent tables.
Page numbers in **Bold** represent figures.

Abu Dhabi military base 45
academic history 37
academics 122
Action Plan 111, 200, 213
activism 50, 109
actors 4–5
added-value sectors 124
Adler, E.: and Crawford, B. 8
administrative institutions 9
Agadir Agreement 158
*Agence Française de Developpement* (FDA) 185
agential coalition 14
agricultural goods 217
agricultural output 175
Algeria 198, 218
Aliboni, R. 20–1
American Free Trade Area 156
Anna Lindh Foundation 159
anti-Americanism 104
Arab activists 213
Arab concurrence 104
Arab Maghreb Union 158
Arab Peace Initiative 98
Arab political leaders 134
Arab polity level 141
Arab regimes 139
Arab Spring 216–17
Arab television networks 42
Arab World Institute 42
Arab-Israeli conflict 97, 102, 139
Arab-Israeli peace process 8
Arafat, Y. 103, 118
architecture 3
Asseburg, M.: and Salem, P. 141
Association Agreements 14, 89
Association Council 119
Association of South East Nations (ASEAN) 159
Aznar, J.-M. 66

Balfour, R. 24; and Schmid, D. 143
Balkan countries 158
Barcelona Conference 5
Barcelona Declaration 28
Barcelona Process 212, 221
Barnier, M. 61
Berlin Declaration 118
Bicchi, F. 1–15, 19
bilateral summits 69
bilateral ties 125
bilateral trade: implementation 175
bilateralism 6–8, 218–20
Blair, T. 102
Border Assistance Mission at the Rafah Crossing Point (EUBAM Rafah) 90
Brussels European Council 86
budgetary issue 70
Bush, G.W. 104
Business Development Initiative 182–3

Chirac, J. 39, 102
Civic Democrats 92
civic protection 184
civil protection 180–1
civil society 27, 68–9; contribution 138
Civil Society Facility for the Neighbourhood 220
civil society organizations 29
civilization 8
climate change agenda 204
Clinton, B. 106
co-governance 215
co-operation 37–52
co-ownership 19–34, 68
co-presidency 23–6

# INDEX

Cold War 13–14, 84
Comelli, M. 147
Common Foreign and Security Policy (CFSP) 5
commonality 82
competences 20
Conclusions of the Council Ministers 107
conflicts 12
continuity 183
converge 140
Crawford, B.: and Adler, E. 8
cross-border co-operation (CBC) 60
cultural diplomacy 40
Curiel, R. 124
Czech National Trade Promotion Agency 91
Czech Republic 87–93

Darbouche, H. 191–206
de Villepin, D. 37–8
de-politicization 140–2
decision-making processes 34
declaratory policy 112
Del Sarto, R.A. 115–29
Delgado, M. 37–52
democracies 147
democratization 104, 219
destabilization 112
development 169–188
Development Banks 174
deviant voices 145
diplomacy 66
diplomatic presence 110
discourse 162
disputes 12
diversification potential 200
domestic constituencies 92
domestic politics 105

Eastern Europe 77–93
Eban, A. 117
economic activity 155
economic focus 176
economic liberalization 172
economic recession 69
economics 124
economies 73
Egypt 11, 105, 110–11
electoral campaigns 43
elites 93
Emerson, M. 143
energy co-operation 191–206
energy dimension 201–2

energy disputes 199
energy policy 194, 199
energy security 191; Green Paper (2000 EU Commission) 198
energy sources 202
energy suppliers 203
enthusiasm 65
Eran, O. 124
EU 34, 50; agenda 4; enterprise 101; foreign policy analysts 137; foreign policy initiatives 115; initiatives towards Southern neighbours **13**; statements 107
EU Commission 187; Green Paper on energy security (2000) 198
EU-Israel Action Plan 123
EU-Israeli differences 121
EU-Israeli relations 117–25
Euro-Arab quarrel 126
Euro-Arab relations 133–48
Euro-Atlantic structures 84
Euro-Israeli diplomatic relations 116
Euro-Med agenda 60
Euro-Med Association Agreements 5
Euro-Med Market programme 159
Euro-Med University (EMUNI) 181–2
Euro-Mediterranean Conferences of Ministers of Foreign Affairs 20
Euro-Mediterranean Economic and Social Committee 162
Euro-Mediterranean Energy Partnership (EMEP) 192
Euro-Mediterranean Information Society (EUMEDIS) 159
Euro-Mediterranean Institute for Risk Science 182
Euro-Mediterranean Investment and Partnership (FEMIP) 160
Euro-Mediterranean Parliamentary Assembly (EMPA) 28
Euro-Mediterranean Regional and Local Assembly (ARLEM) 28
Euro-Mediterranean relations 1–15, 219
Europe 100; actors 73; agenda 140; aid 108; energy security 145–6; financial support 117; foreign policies 134; initiative 110; mainstream 61; normative narrative 105–8; public opinion 122; scholars 138
European Bank for Reconstruction and Development 217
European capitulation 148
European Coal and Steel Community 163

# INDEX

European Commission 26
European Communities (EC) 116
European Council 52
European Endowment for Democracy 220
European Energy and Climate Package 201
European Environment Agency (EEA) 178
European External Action Service (EEAS) 24
European institutions 23
European Investment Bank (EIB) 9, 164, 197, 217
European Neighbourhood Investment Facility 178
European Neighbourhood and Partnership Initiative (ENPI) 173
European Neighbourhood Policy (ENP) 2, 213; bilateralism 6
European Neighbourhood Policy Instrument (ENPI) 136
European Parliament 27
European Space Agency 119
Euskadi Ta Askatasuna (ETA) 61
external relations 70, 155–7

Fayyad, S. 108
Federation of Czech Jewish Communities 88
Federation of German Industries (BDI) 80
Ferrero-Waldner, B. 120
France 2, 109; European allies 47; foreign policy 40, **44**, 63; foreign policy personalization 48; Mediterranean policy 43; political interests 136; unilateralism 62–6
Franco-German axis 38
Franco-Spanish relations 58
free trade area 7
Füle, S. 164, 212
functionalism 9–13, 111
functionalist approach 153–66

Gas Exporting Countries Forum (GECF) 198
*Gaz de France* (GDF) 201
Gaza War 126
General Affairs and External Relations Council 85
geographical regionalism 161
geopolitical preconditions 93
German Chancellery 86

German-Arab Association 80
German-Arab economic forums 80
Germany 4–5, 77–93, 109–10; foreign policy 78–81; foreign policy agendas 78; foreign policy behaviour 83
Giammusso, M. 165
Gillespie, R. 57–73, 211–221
global powers 203
governance structures 219
governments 100
Granada 69
Great Britain 109–10
Grzegorz, P. 84
Guaino, H. 21, 41, 44, 49

harmonization 158
heterogeneity 34
Hill, C. 61
Hispano-French relations 59
Holden, P. 153–66
Hollis, R. 97–112
Horizon 2020 Programme 177–8
human rights 12, 148
Human and Social Development Workshop 182
humanitarian aid 118
Hungary 87–93
Hunt, D. 169–88

implementation visibility 185–6
import liberalization 173
individual actors 108–11
individualism 37–52
institutional form 163
institutional innovation 59
institutional logic 6–13
institutional paralysis 22
institutional parameters 26
institutional structure 170
institutional terms 99
institutionalist literature 22
institutionalization 7
institutions 68, 139–40; building 108; co-operation 156; design 50; entities 27–9; logics 14; overlaps 111–12; reform 185
integration 144, 220
inter-state tension 32
intergovernmental dynamic 50
intergovernmental setting 21
intergovernmental veto-playing 30
Interministerial Mission 41
international context 98
International Court of Justice (ICJ) 107

international donors 121
international factors 148
international frameworks 85
international law 109
international organizations 162
international power relations 192
internationalism 103
intra-European fragmentation 8
Iran-Iraq war 109
Islamist extremism 100
Israel 3, 27; international status 109; invasion 109; political circles 88

Jewish civil society organizations 89
Johansson-Nogués, E. 19–34
Joint Permanent Committee (JPC) 137
Jordan 12, 110–11
Jouyet, J.P. 47

Kausch, K.: and Youngs, R. 146
Kohl, H. 62, 79
Kouchner, B. 66, 214

land highways 178–9
Le Roy, A. 49
leadership 45–7
legislative elections 104
Lequesne, C.: and Rozenberg, O. 49
liberalization 175
liberalization agreement 119
Liberman, A. 127
Libyan Transitional Council 214
Lieberman, A. 31
Lisbon Treaty 15, 31, 65
Livni, T. 118
Lorec, P. 180

mainstream thinking 216
market liberalization 203
Marseille Declaration 24, 27–8
Marxism-Leninism 77
Masa'deh, A. 26, 211
Mediterranean Business Development Initiative (MBDI) 184
Mediterranean Centre for Scientific Research 182
Mediterranean crude oil data *196*
Mediterranean dimension 161
Mediterranean economies statistics *171*
Mediterranean Environmental Technical Assistance Programme (METAP) 177
Mediterranean initiative 51
Mediterranean integration 102, 144
Mediterranean markets 172

Mediterranean natural gas data *195*
Mediterranean Partner Countries (MPCs) 98, 137
Mediterranean Solar Plan (MSP) 179–80
Mediterraneanization 44
Merkel, A. 64
*Mesures d' accompagnement* (MEDA) 158
micro-finance project 184
Middle East diplomacy 64
Middle East Peace Process (MEEP) 11, 90
Middle East Quartet 81
Middle East sensitivities 70
Middle East tension 25
Ministry of Foreign Affairs 25, 42
Ministry of Industry and Change 91
*Mission interministérielle de l'Union pour la Mediterranee* organization chart **41**
modernization 172
modernization theory 142
Monnet, J. 46
Moratinos, M. 66, 214
Morocco 58; rainfall quality *172*
Muburak, H. 127
Multi-fibre Agreement (MFA) 174
multilateral dimension 7
multilateral level 160
multilateral structures 88
Muselier, R. 42

national impetus 59
national influence 72
national interests 82
national investment 180
neo-functionalist theory 157
neo-liberal approach 172
neo-liberal vision 156
neo-liberalism 156
Netanyahu, B. 117
nominal recognition 194
normative agenda 100
North Africa 200
North American Free Trade Area 160
North Atlantic Treaty Organization (NATO) 101

Obama, B. 3
Occupied Palestinian Territories (OPT) 30, 101
operationalization 148
Organization of Petroleum Exporting Countries (OPEC) 193

# INDEX

Oslo Accords 106
Oslo Agreements 11

Palestinian Authority (PA) 85
Palestinian Fatah movement 107
Palestinian National Assembly 26–7
Paris Declaration 25
Paris Summit (2008) 29
Parsley Island 58
partner concerns 20
Partnership for Democracy and Shared Prosperity 215, 216
paternalism 136
peace agreement 126
peace diplomacy 71
peacemaking 121–2
Poland 83–7
Police Co-ordinating Office for Palestinian Police Support (EUPOL COPPS) 90
policy *acquis* 157–60
policy framework 142, 218
policy initiative 87
policy pronouncements 197
policy renewal 221
policy reorientation 38
Polish paradigm 87
political commonalities 140
political context 15
political energy 73
political features 123
political initiative 2
political institution 65
political interests 146
political leadership 31
political perspectives 212
political preferences 145
political profile 79
political reform 141, 147
political regimes 144
political relations 127
politicization 9–13, 23, 111
politics: strategic issues 143
politics agenda 164
pollution control 177
Potuznik, J. 89
power considerations 93
practice and rhetoric: divergence 120–1
pragmatism 7, 144
Prevention, Preparedness and Response to Natural and Man-made Disasters (PPRD) 181
priority objectives 194
project themes 176, 183–7

projects 68
public agencies 157
public-private partnerships 162

Rabin, Y. 117
radical rethink 212–16
rationale 82
realist scenario 144–7
reforms 33
region-building 67–8
Regional Economic and Development Working Group (REDWG) 106
regional integration 173
regional market structures 192
regional politicization 3
Regional Transport Project 178
regionalism 6–8, 154
regulatory area 199
relations 33, 161
renewable energy 179
Renewed Mediterranean Policy 28
rhetoric and practice: divergence 120–1
Riker, W.H. 2
Rozenberg, O.: and Lequesne, C. 49
Russia 199

Salem, P.: and Asseburg, M. 141
Sarkozy, N. 19
Schlumberger, O. 133–48
Schmid, D.: and Balfour, R. 143
Schori, P. 122
Schroeder, G. 79
Schumacher, T. 77–93
Schwarzer, D. 136
*Secrétariat General du Gouvernement* (SGG) 41
security 217
Seeberg, P. 146
Sharon, A. 103, 118
Six Day War (1967) 88
social situation 39
societal attitudes 86
socio-economic areas 154
socio-economic realities 32
socio-political significance 163
solar power 204
Southern Mediterranean Countries (SMCs) 193
Soviet Union 87
Spain 57–73; regional policy 60
structures 33
sub-regional components 179
sub-regional groupings 178
sub-regionalization 38

# INDEX

summits 23
supranationalism 215
symbiosis 97–112
Syria 85, 216
Syrian-Israeli peace negotiations 49
Szejnfeld, A. 85

Taliban regime 103
Telle, S. 49
terrorism 103
Topolánek, M. 89
totalitarianism 83–4
traditional leadership 52
transnational co-ordination 134
transnational projects 153
Tunisia 125, 174, 217
Turkey 10, 99
Tusk, D. 81

unilateralism 39
Union for the Mediterranean (UfM): founding Summit Declaration 176; structure 22–9

United Kingdom (UK) *see* Great Britain
United Nations Development Programme (UNDP) 141
United Nations Environment Programme (UNEP) 177
United Nations Interim force in Lebanon (UNIFIL II) 81

Van Rompuy, H. 69
Venice Declaration 105

Walesa, L. 83
Warsaw Pact 83
Western European Union (WEP) 101
Western Europe 38
Work Programme 70
World Bank 134
World War II 14

Youngs, R.: and Kausch, K. 146